P9-CCX-507

The Little Know-It-All

Common Sense for Designers

die gestalten verlag

1.0

Over the last few years, the designer's role and job description have under-gone rapid and substantial change, expanding to encompass a huge range of new media, tools, resources – and, consequently, responsibilities.

Where specialists once dealt with anything from fair drafting to post production, while dedicated legal departments would take care of all contract or copyright issues, today's designers must do more themselves. Clients want more than just the creative implementation of their ideas; they now hire designers for their specific visual styles, which they expect to see translated across a range of different media. Thus, today's designer requires in-depth knowledge of all aspects of the design process, of the potential pitfalls, how his or her work links in with other disciplines, and – not least – how to sell the work and skills to his or her best advantage. Even with the best of train-ing, this is an ambitious requirement.

Today's design graduates, however, often lack the necessary theoretical foundation and hands-on experience to cope with the myriad issues outside of their immediate realm. And considering the breadth of knowledge required to shine in a job that touches on so many different facets of the creative pro-fession, even the most veteran designer will eventually find himself or herself confronted with the great unknown.

In view of the wide array of specialist literature available, there is a con-spicuous lack of concise yet comprehensive works on the essential issues today's designer is likely to face – from typography to production, from digital media to legal concerns. While internet research provides an often only long-winded and sometimes questionable alternative, this is where The Little Know-It-All, our quick, no nonsense and very much up-to-date refer-ence work, comes into play. Brimming with essential facts, this little helper is always on hand to fill each designer's individual knowledge gap with con-centrated information.

Foreword

Accessible and well structured, this handy book is a contextualised manual that presents the big picture as a pared down, yet comprehensive volume.

Divided into the chapters "Design", "Typography", "Digital Media", "Production", "Marketing", "Law" and "Organisation", and interspersed with illustrations to visualise the terms discussed, the book offers concise descriptions and evaluations of past, present and future standards as well as explanations of the field's unique nomenclature. An interdisciplinary collection of all the facts a designer needs to know (with a few more thrown in for good measure), which are cross-linked and embedded in both a historical and practical context, the compact tome covers a wide array of topics from kerning to typographical measuring units, from organisational guidelines to copyright law, from paper types and coats to colour profiles and printing techniques. And in view of its global audience, the book also touches on different formats and the idiosyncrasies of international law.

While those in a hurry will appreciate the index featuring more than 2,000 highlighted keywords and technical terms, this little jack-of-all-trades works equally well read cover to cover. At the end of each chapter, ambitious souls in need of more in-depth information will find a selection of links and references to specialist sources. There's also plenty of room for notes here and in a special section at the end of the book. So, whether you start at the beginning, pick a term from the index, jump from reference to reference or browse for the odd inspiration, the Little Know-It-All acquaints you with all the right shortcuts from theory to practice and makes an excellent fallback for anyone in need of quick, no-nonsense information.

Compiled by designers for designers, this mix of design common sense, hands-on expert knowledge and tips from the profession's luminaries is destined to become a fixture on your desk in no time at all.

The Editors

2.0

Contents

1.0	Foreword	2
2.0	Contents	5
3.0	Design	6
4.0	Typography	54
5.0	Digital Media	100
6.0	Production	150
7.0	Marketing	216
8.0	Law	242
9.0	Organisation	282
10.0	References	308
11.0	Notes	314
12.0	Index	324
13.0	Imprint	350

3.0

Welcome to the golden rules of design – and a few pointers on when and how to break them.

Whether the style is clean-cut and simple or out-there experimental, whether the project is a hand-sketched record sleeve or a polished corporate identity – great designers tend to draw on their ability for lateral thinking, for unusual combinations, and for bringing a brand new take even to visuals that were among our trusty old favourites. And yet, in order to turn these fleeting flashes of genius into actual, working templates, we require more than just the right tools of the trade and a firm grasp of the necessary skills: what we need is a solid grounding in the fundamental principles of design.

In addition to essential elements such as light and colour, perspective and space, signs and symbols, this chapter also sheds light on the psychological twists of subjective perception that send our synapses into a spin – how to use them to our advantage, conjure up meaning between the lines and reach our very own vanishing point.

So don't fret – in-depth knowledge and understanding of these ground rules will not automatically confine your work and spirit to a strict aesthetic framework, but simply help you fine-tune your inner eye to shape dot, line and area into something entirely your own.

Design

3.1	**Light and colour**	8
	Colour theories	10
	Colour effects	11
	Colours on the web	14
3.2	**Perspective and Spatial Quality**	15
	Three-dimensional display	16
	History of perspective	17
	Perspective processes	18
	Parallel perspective – parallel projection processes	21
3.3	**Illusions**	24
3.4	**Fundamentals of Design**	25
	Basic elements	25
	Contrasts	28
	Rhythm, symmetry and asymmetry	30
	Laws of design	31
	Figure-ground distinction	33
3.5	**Signs and Symbols**	34
	Types of signs	35
	The logo	37
3.6	**Layout and Page Design**	41
	Layout process	41
	The Golden Section and harmonious page formats	42
	The design grid	42
3.7	**Creativity**	44
	Creative techniques	44
3.8	**Golden Rules of Design**	48
3.9	**Tips and Links**	49

Light and Colour

Heraclitus said that "seeing is a delusion" a very long time ago. In fact light is colourless. **Perceiving colour** is simply a human sensory experience, usually caused by a physical stimulus. This external colour stimulus is projected on to the inside of the eye via the cornea, the pupil, the lens and the vitreous body, finally reaching the **retina** – a complex layer, consisting essentially of nerve endings. Here the nerve cells are stimulated. The **colour stimulus** can come either directly from the light source or indirectly, via the surface of a solid.

In terms of physics, **light** is electromagnetic radiation. These energy beams are **electromagnetic waves** that differ in wavelength. The human eye is sensitive to only a very limited range of this electromagnetic spectrum. Its sensitivity to colour varies according to the wavelength and intensity of the electromagnetic radiation in the visible range. The eye can perceive wavelengths of roughly 400 to about 700 nanometres (nm). The average viewer can distinguish between 100,000 and 1,000,000 nuances of colour. If white light falls on a glass prism through a narrow crack, the light is refracted and the so-called **colours of the spectrum** (fig. 1) become visible: red, orange, yellow, green, blue, indigo and violet. These encompass the range of visible light. Ultraviolet radiation (UV radiation) lies below 380 nm, and infrared radiation (IR radiation) above 780 nm; human beings can also perceive the latter as heat. Short waves are X-rays or gamma rays, for example.

Let us look at the retina again. On its front surface are **light-sensitive sensory cells**, **receptor cells** that look like **rods** or **cones**. These rods and cones sit next to each other as if in a little box and are sensitive to light in various areas of the spectrum.

1

Wavelength

Ultraviolet   Visual radiation (light)   Infrared

According to Küppers,[1] each "receptor type" is allotted a certain "receptor power": the rods are responsible for seeing in low light (different shades of light and dark can be perceived), and the cones make it possible to see in daylight and to distinguish colours. The incidental light triggers chemical signals in the eyes, and these are transformed into electrical stimuli, which are then processed in the brain. The brain then compiles the image it has perceived from the wide range of detailed information.

Among the cones, three different receptors handle red, green and blue. Colour vision results from the superimposition of these three colours. For example if red and green rays of light impact on the appropriate receptor cells in the eye, this produces the compound colour yellow; the human eye's limited powers of resolution mean that the colours mingle to form a colour impression. If all three colour receptors (red, green and blue) are stimulated, then the eye will see a white image – this is the principle of **additive colour mixing**.

Additive colours (light colours) are created by light addition. In theory, all the visible colours can be formed with the additive primary colours (→ pp. 10, 11) **red**, **green** and **blue** (RGB), and all three of these colours when mixed equally produce white – if projected on top of each other. If one colour is missing, black is produced. Additive colour mixing is used in colour monitors, for example (→ ch. Production, p. 169, RGB). Mixing two of these colours in equal proportions produces the **secondary colours** (→ p. 11) **cyan** (C), **magenta** (M) and **yellow** (Y), which in their turn make up the basic colours in the **subtractive colour** system (→ ch. Production, pp. 169, 186, CMYK). This colour system describes the behaviour of **non-luminous colours**. The subtractive process is based on the **absorption** of the colours of the visible light spectrum of white light.

---

[1] Küppers, Harald: Basic Law of Color Theory (1981)

Additive colour mixing: red + green = yellow; green + blue = cyan; blue + red = magenta; red + green + blue = white (fig. 2).

Subtractive colour mixing: cyan + magenta = blue; magenta + yellow = red; cyan + yellow = green; cyan + magenta + yellow = black (fig. 3).

The colour impression a viewer gains of an object is created by the light component that meets the surface of the object and is reflected thereby, while the remainder is absorbed on impact. Here, the reflected light determines the colour of the object. An object that does not reflect any of the component colours of sunlight therefore always looks black. Conversely, an object that reflects all the colours is seen as white. **Non-luminous colours** are also called **sub-traction colours** because of absorption.

Light temperature. If the temperature of a glowing metal body is gradually raised it radiates different colours of light. The light colour temperatures are derived from this: **warm light**: 3500 K (Kelvin)[2], **daylight**: 5500 K, **cold light**: 6500 K.

Colour theories

Important figures from the fields of culture, art and science such as Isaac Newton and Johann Wolfgang von Goethe, as well as Paul Renner, Johannes Itten and Harald Küppers, have addressed the phenomenon of colour in detail. Several theories and doctrines still retain their validity. Two of the most important colour theories are explained below.

The Swiss painter and art historian **Johannes Itten** (1888–1967) worked on the basis of **three basic colours**, the primary colours **blue**, **yellow** and **red**, which produce the **twelve part chromatic circle** (1961) when mixed (fig. 4). The three basic colours of the first order are found in the middle of the colour circle. Itten largely based his theory on the insights gained by the poet and natural historian Johann Wolfgang von Goethe (1749–1832).

---

[2] Boltzmann constant of average kinetic energy, value: 1.380658×10^{-23} joules per kelvin (Ludwig Boltzmann, physicist, 1844–1906)

According to Itten, **complementary colours** are two colours that lie opposite each other in a chromatic circle – in other words, which have complementary positions to each other, like for example red and green (fig. 5). Mixing two complementary colours produces grey. According to Goethe and Itten, the colour combination of two complementary colours is seen as "especially harmonious".

The colour expert Harald Küppers, internationally acknowledged in the printing industry and the graphic arts trade, bases his colour theory on six basic colours: yellow, green, cyan-blue, violet-blue, magenta-red and orange-red. He arranges the basic colours in a hexagon, rather than a circle (fig. 6).

The elementary colours (violet-blue, green, orange-red) and the basic colours according to Harald Küppers are explained below.

In Küppers' theory, primary colours are the starting point for any colour process. In subtractive mixing ("SubMi") these are the basic chromatic colours yellow, magenta-red and cyan-blue, and their interaction with the background colour white. In additive mixing the primary colours are the basic chromatic colours orange-red, green and violet-blue, and their interaction with the background colour black. According to Küppers, secondary colours are created by mixing two primary colours, and tertiary colours are created by mixing three primary colours.

Colour effects

Colours produce their effects in a variety of ways. **Colour psychology** and **colour symbolism** offer various interpretations of individual colours. In fact, colours and their meaning, or the sensations they produce, are not solely dependent on individual experience, but also on agreed impressions that go back over the centuries. So colours can mean different things in different cultures, and the following list of colour interpretations should not be regarded as definitive.

White (fig. 7) is the sum of all colours in terms of additive colour mixing (→ pp. 9, 10). It is perfect, and symbolises light, faith, the ideal, the good, the beginning, the new, cleanliness, purity, innocence, modesty, truth, neutrality, intelligence, science and precision – but it also stands for emptiness and the unknown. Anything intended to be hygienic is also white. Being "whiter than white" suggests irreproachable behaviour. The word "candidate" comes from the Latin "candidus" and means gleaming white, snow-white, natural, straightforward and sincere. White is not a colour; it is **achromatic**.

Yellow (fig. 8) stands for sun, light, ripeness, warmth, clarity, planning, law, optimism, thrusting forwards, sensitivity, luxury, joie de vivre, freshness, cheerfulness, kindness, change and extroversion – but it also suggests conceit, presumption, arrogance, envy, jealousy, miserliness, egoism, lies, uncertainty, lack of feeling, as well as bitterness and venom. In English, "yellow" also means cowardly. In history, yellow was always the colour of outlaws, the persecuted and the rejected.

Orange (fig. 9) symbolises joy, efficiency, liveliness, fun, extroversion, excitement, affirmation of life, exuberance, energy, activity – but also seems fanatical and symbolises roughness and pushiness.

Magenta (fig. 10) stands for idealism, gratitude, commitment, order and sympathy – but also for snobbishness, arrogance and dominance.

Red (fig. 11) is vitality, joy, activity, activation, energy, dynamism, temperament, impulsiveness, warmth, passion, temptation, arousal, the will to conquer, zest for action, eccentricity and vigour – but also strain, danger, violence, chaos, anger and hatred. Fire is the symbol of the divine and holy, but also of blood and war.

Violet (fig. 12) signifies mysteriousness, magic, vanity, extravagance – but also pushiness and conflict.

Blue (fig. 13) symbolises relaxation, outcome, results, sympathy, trust, friendliness, reliability, expansiveness, recreation, harmony, contentment, calm, silence, passivity, infinity, cleanliness, hope, consolidation, cleverness, science, preservation, longing, imagination, courage, sportiness – but it can also convey cold, coolness, dreaminess, neglectfulness, obstinacy, naïveté, satiation or melancholy. Blue is the

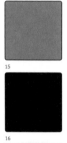

coldest colour of the spectrum and the complementary colour
(→ p. 11, 29) to orange. Blue is also the colour of distance. The
warmer it is, the closer a colour seems; the colder it is, the
further away a colour seems (→ p. 19, Colour perspective).

Green (fig. 14) symbolises stamina, freshness, tenacity,
relaxation, nature, vegetation, naturalness, calm, generosity,
health, confidence; it seems outgoing, joyous, regenerating,
complete, neutral – but also suggests indecision, laziness
and impersonal behaviour.

Black (fig. 15) is resistance, impenetrability, darkness,
functionality; it seems dramatic, mysterious, reserved –
but also symbolises heaviness, brooding, gloom, negation,
isolation, constriction, pessimism and hopelessness. Black
is not a colour; it is **achromatic**.

Grey (fig. 16) indicates neutrality, reflectiveness, serious-
ness, objectivity, functionality, plainness, readiness to compromise – but also negativity, uncertainty, coldness, lack of
involvement, inconsolability and misery.

Checklist: Choosing a colour climate

– What message and basic atmosphere are to be conveyed?
– What target groups are to be addressed?
  (Pay attention to cultural and age-specific differences.)
– What contrasts and harmonies should be produced?
– Which basic colours should be used?

Note: As a rule, not more than four basic colours should be
used. A few well-matching colours are more effective than a
lot of different colours.

To make the information more accessible to users, the
same topics or subject matter should always be presented in
the same colour. It is possible to work with shades of colour
within a particular topic.

Different content should be clearly identified with distinguishable colours. Passages that need to be emphasised
(such as mnemonics, hints, names etc.) can be flagged with
a colour contrast (→ pp. 28, 29), in the form of a highlight
(→ ch. Typography, p. 78, Accentuation).

Colours on the web

Colours are displayed in the RGB colour mode on monitors, so websites use only **RGB colours**. RGB values are usually given in hexadecimal[3] form.

The numerals 0–9 as well as the letters A–F are available for this purpose. Thus, for example, the RGB value "white", which contains 255 parts red, 255 parts green and 255 parts blue, is written decimally as 255, 255, 255. In hexadecimal form white is written as #FFFFFF. The # indicates that this is a hexadecimal value.

Web standards ensure that users can have access to the same web-compatible colours. The first web standards were developed at a time when computers could often display only 256 different colours (**8 bit**). This led to the development of a **web-safe palette** of **216 colours** that left out the colours that are displayed differently on different operating systems.

If images with greater colour depth were used, the colours that could not be displayed had to be simulated, for example by **dithering**, a technique that deals with missing colours with a particular pixel arrangement using the available colours. This effect was avoided by using the **standard palette** with 216 colours.

For a time, 22 of the 216 web-safe colours were raised to the status of so-called "really web-safe colours" that could be reproduced correctly in 16 bit colour depth without being recalculated.[4] As graphics cards and monitors that could work with colour depths of **24 bits** or more became commonplace, this range became less important too. Today, all the colours in the hexadecimal system can be used. But colour reproduction still depends on the physical qualities of the display devices and the user settings chosen, such as brightness and contrast.

---

[3] Hexadecimal system (Greek: hex = 6; Latin: decem = 10; decimus = tenth; sexadecimal system = 16): number system on base 16.
[4] Stern, Hadley; Lehn, David: Death of the Web-Safe Color Palette (2000)

Perspective and Spatial Quality

"The natures of perspective are three in number. The first sets down why things that are moving away appear smaller, the second propounds the knowledge whereby colours change and the third and last sets down the explanation of why the outlines of things are to be represented less precisely," said the universal genius Leonardo da Vinci (1452–1519) on the subject of perspective. The Dutch graphic artist Maurits Cornelis Escher (1898–1972) also went into some detail about the strange tension implicit in any reproduction of a three-dimensional situation on a flat surface. He contradicted a viewer as follows: "Drawing is illusion; it suggests three dimensions although only two are there. And however great an effort I make to persuade you that it is an illusion, you will still continue to see three-dimensional objects."[5]

Perspective is the **constructed** or **pictorial representation** of an object or space on a **two-dimensional surface** – using resources that convey an impression of **spatial depth** or solid quality. Perspective can be represented with perspective lines and lines indicating space. The Italian artists of the Renaissance applied the word "**prospektiva**" (vista) to convey an optical impression of three-dimensional quality on the basis of a geometrical system.

Three-dimensional vision is the result of **binocularity** – seeing with two eyes (fig. 17). We can see in three dimensions because our two eyes see the same image from different angles. It is only in the brain that the two images combine to create a three-dimensional whole – the image.

17

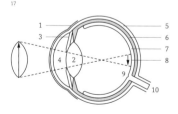

1	Cornea	6	Choroid
2	Lens	7	Retina
3	Iris	8	Fovea
4	Pupil	9	Optic disc
5	Sclera	10	Optic nerve

Ernst, Bruno: The Magic Mirror of M.C. Escher (1987)

Three-dimensional display

In graphic design, three-dimensional images can be created using various creative tools. As a rule, several of them are used in combination.

Light and shadow. The three-dimensional, plastic impression made by an object or solid is created by light and shadow. Shadow is created wherever there is a **light source**, an **object** and a **projection area**. Here the nature of the light source (e.g. focused light, more general light) and its position in relation to the object are the keys to the three-dimensional situation.

One particular task performed by shadow in design is to imitate reality effectively (compare fig. 18 and fig. 19). But images used in **comics** and also **illustrations** can show that a shadow does not always correspond with reality.

Types of shadow. The **object shadow** (fig. 20) is the shadow cast by an object itself on the side that is not facing the light. The **hard shadow** or umbra is the darkest part of a shadow. The **soft shadow** or penumbra is the name for the area of partial shadow between regions of the hard shadow and full illumination (fig. 21).

Tip: *If they are to represent three-dimensional quality correctly, designers should note that the lighting in a represented space is the same for every object in that space. As soon as the actual level and direction of a light source has been established, it is simple to determine the position, shape and size of the soft shadow on a surface. Anyone with a practised eye and knowledge of the rules can draw soft shadows freehand.*

In creative design, the **three-dimensional effect** can be achieved by differences in size (fig. 22), overlapping elements, nuances of light and dark colour, colour shading and differences in focus. Almost any tools and design techniques can be used, e.g. pencil, charcoal, paintbrush, dye, spray paint, printing techniques and the computer.

Collage should also be mentioned at this stage as a design technique. Montage does not just mean dissecting images and reassembling their components to make a new image, using a particular principle for the arrangement. Particular pictorial details can also be enhanced by this method. Conse-

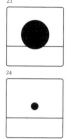

23

24

quently, this technique can rearrange various elements to create a new whole, which can even change the statement made by the original image. A photograph can be made to express something different by drawing or painting on it, and entire motifs can be alienated and placed in a new context.

Size differences. If three-dimensional situations are depicted in a way that is true to nature in a drawing, the scale is reduced as the distance from the observer increases. More distant objects look correspondingly smaller (figs. 23, 24).

History of perspective

The **ancient** world was familiar with the technique of illustrating the **size of individual objects** by shortening depth lines, and created the first illusionistic effects in their stage sets. But antiquity was not aware of the horizon, and there was no overall concept of perspective. Early Christian and medieval painting used perspective almost entirely to indicate status (→ p. 23, significance perspective).

Naturalistic rooms and landscapes were represented for the first time in Biblical pictures of the early **Italian Renaissance**. The laws of perspective were consistently heeded and applied by architects and painters from the mid-14th century. The **development of perspective** in the Renaissance was closely linked with important scientific discoveries and research. Geometry established clear rules about natural phenomena. The search for **regularity**, for **symmetry**, **harmony** and **rhythm** led to the first theoretical groundings for perspective.

The first written notes were made by the Genoa-born humanist, theoretician and universal scholar **Leon Battista Alberti**[6] in the year 1435. In his three "Books on Painting", he compares the picture with a window that reproduces an apparently real space for us. Here, the eye is a point, and the picture surface becomes a plane section through the visual pyramid. Alberti also developed **Alberti's Frame**, a square wooden frame with horizontal and vertical threads stretched across it at regular intervals. This made it possible to transfer a spatial situation that had been analysed visually onto the drawing board.

6   The first major architectural theorist of the Renaissance (1404–1472).

But the Florentine master builder and sculptor **Filippo Brunelleschi**[7] is considered to be the actual father of modern perspective. He tried to prove that the picture as a work of art could also achieve the perfection of applied mathematics. He is said to have used the ideas of the ground plan (→ p. 20) and elevation to this end.

The German painter and copperplate engraver **Albrecht Dürer** was one of the most versatile personalities of art history. He also explained the process of **perspective construction** easily and comprehensibly with a mechanical drawing device.

This all led to the concept of central perspective with **one horizon** and **one vanishing point** for all orthogonal lines running vertically to the plane of the picture.

Leonardo da Vinci was the first to describe **central perspective**. He equated seeing with cognition even in his treatise on painting, and required artists not just to present a realistic or decorative reproduction of nature, but also actually to understand scientific forms and laws. The artist should translate these into images using insights about optics and geometrical processes. Da Vinci's mathematical calculations formed the basis of his detailed studies of the anatomy of the eye.

25

Perspective processes

Linear perspective represents objects by drawing outlines, contours and borders using only the resources of draughtsmanship – without shades of light and dark, hatching or colours (fig. 25). Technical sketches, designs or plans are usually prepared as line drawings.

In **central perspective**, parallel edges are not presented as parallel, but meet at an imaginary point – the so-called **vanishing point**. Central perspective is a projection process with **one vanishing point**.

⁷ Buildings include the dome of the Duomo in Florence, the church of San Lorenzo, and the Pazzi chapel in the monastery of Santa Croce (1377–1466).

26

27

28

29

30

Golden rules of central perspective

1. The front view of an object, in other words the surfaces facing the viewer, is depicted with its correct proportions and angles.
2. The object is placed parallel with the plane of the picture.
3. All lines vertical to the picture plane representing the depth of the object meet at a vanishing point.
4. All lines parallel with the picture plane are foreshortened.
5. The lines placed parallel with the picture plane remain parallel in height and width.
6. All the angles formed by the lines placed vertically to the picture plane change.
7. The angles at the front of the object remain the same (fig. 26).

As the horizon line necessarily shifts up or down in relation to a changing eye level, the extreme versions of central perspective are called **worm's-eye view** (from below) and **bird's-eye view** (from above). Figs. 27 (worm's-eye view) and 28 (bird's-eye view) illustrate the different eye levels and the corresponding effects in a space presented in central perspective (view in "normal perspective", fig. 29).

Other perspective processes include images with **two vanishing points – diagonal perspective –** and also perspective with **three vanishing points**. In contrast to central perspective, perspective with two vanishing points places the object diagonally to the picture plane (fig. 30).

Colour perspective. "The eye, as it cannot move, will never grasp the distance that lies between one object and another through linear perspective alone, but only through colour perspective," said Leonardo da Vinci.
    In normal light and weather conditions, objects and landscapes that are far away will seem lighter in colour than those close by. The light is scattered by the intervening layers of air, dust particles etc. It is possible to create an impression of depth with the aid of **colour perspective** by using different shades of colour in the foreground, the central section and the background.

This effect can be achieved because so-called **cold** and **warm colours** create different spatial effects: warm colours thrust forwards, whereas cold colours seem to hold back. Thus the foreground is usually dominated by colours such as yellow, orange or brown; shades of blue, for example, dominate the background.

In **aerial perspective** an impression of depth is created by the fact that contrasts diminish from front to back and brightness increases from front to back. Aerial perspective is closely linked with the idea of colour perspective, as it derives from the same cause.

31

32

Tip: Blue and violet are used to lighten distant objects as the shorter light waves of the colour spectrum are less deflected in the ultraviolet range than in the red range.

As a rule, a room is represented as a **ground plan** (fig. 31), as a **sheer plan** (fig. 32) and in perspective.

Eye level. The horizon shifts upwards or downwards (→ p. 19, worm's-eye and bird's-eye view) according to the viewer's eye level. The **viewpoint** is the point at the viewer's eye level, vertically above the **standpoint** – set, as a rule, at 1.6 m (fig. 33). The eyes focus on the **picture plane** between the viewpoint and the object. This is where the object is depicted. The perspective distances are established by the diagonals, thus fixing the **distance point**.

Here, the **distance** is the presumed distance between the viewer's eye and the picture plane. The edges of an object running into the depth of the picture or the corners of a room are fixed by the **vanishing lines**. In central perspective (→ pp. 18, 19) all the lines meet at a vanishing point. The picture plane is at right angles to the **base area** on the **base line**. The **horizon line** is the line on the picture plane at the level of the viewpoint. The lines of sight are imaginary lines between the viewpoint and the points on the object. The **standpoint** is the point adopted on the base area at which the viewer stands.

33

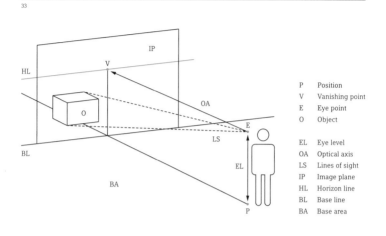

P	Position
V	Vanishing point
E	Eye point
O	Object
EL	Eye level
OA	Optical axis
LS	Lines of sight
IP	Image plane
HL	Horizon line
BL	Base line
BA	Base area

34

Parallel perspective – parallel projection process

Parallel perspective is the perspective view of a solid or space without vanishing points. All the parallel lines in a parallel perspective drawing also run parallel to each other in the image as depicted (fig. 34).

One side of the solid is usually drawn from the front and true to scale. The other sides are at an angle of 45 or 30 degrees, and drawn foreshortened by half. As the vanishing lines run parallel and do not meet at a vanishing point in parallel perspective, all objects appear to be the same size. For this reason, this kind of perspective is not used for depicting landscapes or natural spaces.

Process	Characteristics	Advantages and disadvantages
	Military perspective – The ground plan remains undistorted. – Angle is as wished. – The height lines can be shortened.	– Strong top view. – Also suitable for objects with a non-right-angled ground plan. – Especially suitable for complex objects and buildings.
	Isometric perspective – All surfaces are distorted. – No edges are shortened. – All angles are changed.	– Strong top view. – Most frequent form for technical drawings. – Little three-dimensional effect. – Intelligible dimensions. – Not suitable for centrally symmetrical ground plans (e.g. cubes).
	Diametric perspective – All surfaces are distorted. – All angles are changed. – The depth lines are foreshortened. – Other edges are not shortened.	– Suitable for an object with little depth. – Visual clarity.
	Cavalier perspective – The elevation is not distorted. – Depression angles can be different (e.g. 60 degrees). – Depth lines are halved.	– Simple representation method. – Good visual clarity. – Little effect of depth.

35

Other types

Fisheye projection. The lines that do not run through the centre are curved; areas at the edges are depicted smaller than those in the centre; the viewing angle is 180 degrees and more. Fisheye projection is a distorting mode of representation.

When the **eye** sees in three dimensions (both eyes), it can encompass a span of about 120 degrees. But human beings can only see in really sharp focus over an **angle** of about 1.5 degrees (fig. 35).

In **significance perspective**, the size of the figures and objects represented depends on their importance in the picture, but not on the spatial or geometrical situation. For example, in antiquity and the Middle Ages, figures were arranged in a picture according to their importance or rank: sacred figures are shown large and in the centre, while whole other people or objects are reduced in size, without any consideration of spatial connections or proportions.

Checklist: Perspective construction

The following points should be considered in perspective construction:

- The viewer's standpoint.
- The viewer's eye level.
- The viewer's position in relation to the horizon.
- The positions of the objects in the space.
  Note: The further away an object is, the smaller it appears to the viewer.
- The perspective process.

36

37

38

39

40

41

42

Illusions

There are some geometrical figures that a viewer perceives to be different from how they actually are. For example, lines or distances of the same length can be made to seem different, as F. C. Mueller-Lyer's figure shows (fig. 36). Here, the arrows at the end of the lines create an illusion.

Psychologists have been trying to establish the cause of optical illusions for centuries. One dictionary of psychology defines an **optical illusion** as the "lack of agreement between objective-physical stimuli and sensations or perceptual judgements when assessing the state, … size, spatial position and direction of a stimulus or certain sections of it."[8] Barbara Gillam, Professor of Psychology at the University of New South Wales, has summed up illusion in three points. Her first point is that the illusion is not a product of thought but of perception; even when one knows that one's impression is false, the optical illusion does not disappear (although it can appear weakened if looked at repeatedly within a short period of time). Gillam's second point is that the illusion does not occur through procedures in the retina. And her third point is that eye movements are not involved in creating of the illusion. Other figures that create optical illusions are described below.

For example, in the American psychologist Edward Bradford Titchener's[9] figures, the left-hand inner circle seems smaller than the right-hand inner circle (fig. 37). Zöllner's figures make parallel lines seem to slope in relation to each other (fig. 38). In figs. 39 and 40 all four bars are the same length. But in fig. 40 the horizontal bar seems shorter than the vertical one. The impossible figure (fig. 41), often used as an example in perceptual psychology, confuses the centre of sight by using a skilful combination to make two-dimensional elements placed on the paper as recumbent figure seem three-dimensional (→ p. 33, ambiguous patterns).

8   Fröhlich, W. D.: Wörterbuch Psychologie (2000) No English version available.
9   Conducted studies at Cornell University, Ithaca in order to describe the structure of the mind.

Fundamentals of Design

Basic elements

As a rule, the basic structure of graphic design consists of the basic elements **dot**, **line** and **area**.

The **dot** or point is the smallest element in graphic design. Geometrically speaking, the dot is a zero-dimensional object of zero extent. Geometrically, a point is represented in a flat vector model by X-Y coordinates and in a grid model as an individual cell (→ ch. Production, p. 173, dot).

Depending on distance, points of various sizes can be perceived. Or a grid can be created by arranging points over an area (→ ch. Production, pp. 172–174). This means the points are no longer isolated, but perceived as a grey area (fig. 42).

Designing with dots or points can create a wide variety of visual effects. People start to make limited associations with the mere positioning of a dot at the top, bottom or in the middle of an area. Thus a single point in the centre of an area can convey calm (fig. 43), but trigger tension if it is shifted (fig. 44), create structure and textures by repetition (fig. 45) or admit stimulating and vivid effects through controlled variations in size or colour (fig. 46).

The line. The eye sees an arrangement of dots with a constant distance between them as a line (→ pp. 31–33, Laws of design). Adrian Frutiger, who developed the Univers typeface, said "every linear expression derives from a point set in motion."[10] Unlike the point, which is tied centrally, the line can be dynamic in character. The simplest form of the line is the **straight line**. As a vertical, it makes an active and light effect (fig. 47), as a horizontal a passive and heavy one (fig. 48). Straight lines can also run diagonally (fig. 49). If a straight line runs from bottom right to top right it is read as ascending. But if it runs from bottom right to top left it is seen as a line dropping from left to right. In Western cultures the eye explores an area or shape from left to right because reading habits (→ ch. Typography, p. 61) influence sight. But reading habits can vary by country, as for example in China.

[10] Frutiger, Adrian: Signs and Symbols (1998)

50

51

52

53

54

55

56

Lines can also be bent and curved (fig. 50). They can be used in a variety of ways. They can connect (fig. 51), determine the outlines of forms by the way they run (free lines), structure areas, condense into structures or **hatching** (fig. 52), support and emphasize, or point something out such as a direction. An **arrow** can also make various statements: point straight ahead, towards top right or bottom right. A "turning arrow" is curved and points back in the starting direction (u-turn).

A **stroke** is usually a short, irregular line, essentially the artistic, individual treatment brought about by the way a paintbrush is handled.

A **spiral** is an open line – connected from outside to inside or inside to outside, with a tendency towards strong movement.

Note: The eye perceives a line of type, or other line, placed in the mathematical centre of an area as being too low (fig. 53). Hence a distinction is made between the optical and the mathematical centre. The optical centre is the visually satisfying proportion in design. Bars placed horizontally always look somewhat heavier than vertical ones, as the bars in figures 54 and 55 show. The two bars are equal in length and width (→ ch. Typography, p. 86, diagonal and bold upstrokes).

An **area** is a closed, two-dimensional figure. It is enclosed by a homogeneous surface that is usually presented in two dimensions, and formally limited by one or several linear segments. Geometrically, an area is usually defined by a sequence of pairs of co-ordinates. Simple design devices can create various effects on an area.

A **circle** has no starting and no endpoint, and is therefore a symbol of infinity. It is a two-dimensional figure. Geometrically speaking, a circle is a line whose points are equidistant from the centre. Closing this curved line creates a circle. With the exception of the straight line, the circle is the only flat curve with a constant curvature. It conveys less tension than a rectangle or triangle, as it is not pulling in any particular direction. A circle seems harmonious, complete in itself and infinite (fig. 56).

57

58

59

60

61

62

63

Albert Kapr and Walter Schiller write: "The circle is the sign of balance and repose, the eye is drawn by the centre, which is not identified and yet known."[11]

In the late 4th century BC, Babylonian astronomers and mathematicians used a little circle as zero to identify units missing in certain ordering systems. Later it was written simply as a large dot.

Geometrically speaking, the **ellipse** is a particular, closed curve of oval shape. It seems more dynamic than the circle, for example. Upright it seems to be striving upwards, but also unstable (fig. 57); placed horizontally it seems more in repose (fig. 58).

In geometry the **square** is a regular, two-dimensional rectangle: all four sides are the same length and all four (interior) angles are equal (90 degrees). If the square is standing on one of its sides it seems stable, functional, static and calm (fig. 59). But if the square stands on one of its points, its effect changes (fig. 60).

Geometrically speaking, the **rectangle** is a four-sided figure whose interior angles are all right angles. The opposite sides are parallel and the same length. It can be placed vertically or horizontally. Almost all paper formats are rectangular. The lengths of a rectangle's sides must differ clearly if it is not to be confused with a square. It looks more active than the square. Set horizontally, it seems stable, secure, supporting, heavy, inert and wide (fig. 61). Set vertically, it seems to strive upwards; it is active, light and narrow (fig. 62).

The **triangle** (fig. 63) has a strong directional component. "A triangle with its acute angles seems dynamic, and the eye will be caught by the most acute angle and then seek further in this direction," say Kapr and Schiller.[12] Mathematically speaking, the triangle is a two-dimensional geometrical figure limited by the straight lines connecting three points (A, B, C).

[11] Kapr, Albert, Schiller, Walter: Gestalt und Funktion der Typographie (1977). No English version available.
[12] Kapr, Albert, Schiller, Walter: Gestalt und Funktion der Typographie (1977). No English version available.

Contrasts

Contrasts (Latin: contrastare = against, stand) are produced by extreme opposites. Artistic design devices can include formal contrasts, fore- or background contrasts and contrasts in brightness or **colour**. A state of tension can even be achieved by contrasting "unprinted" and "printed" areas. The following are some of the contrasts that can be created: small-big, heavy-light, light-dark, few-many, closed-open, fine-coarse, matt-glossy, simple-complicated, near-far, dynamic-static, calm-lively, positive-negative, harmony-disharmony, long-short, thick-thin, empty-full, soft-hard, straight-curved, ordered-chaotic, cold-warm, passive-active, or round-angular.

    The contrast control on a monitor produces the desired brightness in an image. The contrast should be set so that the image is sufficiently bright but not manipulated unduly. Excessive manipulation causes the right-hand edges of the image to blur in respect of colour.

Colour contrasts

Johannes Itten put it like this: "We speak of contrast when clear distinctions or intervals can be discerned between two colour effects that are to be compared."[13] He identifies the following seven colour contrasts to this end: colour-as-such contrast, light-dark contrast, cold-warm contrast, complementary contrast, simultaneous contrast, quality contrast and quantity contrast. All of these colour contrasts are based on human perception and sensation.

Colour-as-such contrast – also known as **chromatic contrast** – is created automatically as soon as colours are used (fig. 64). It also identifies the contrast between two different colours.

    Light-dark contrast. White and black (fig. 65) create the strongest light-dark contrast, violet and black the smallest contrast.

    Warm-cold contrast (fig. 66). The colours in the left-hand half of the spectrum – from blue-violet to yellow-green – are the so-called **cold colours** (→ p. 20). The colours in the right-

64

65

66

67

[13] Itten, Johannes: The Art of Color (1997)

hand half – yellow to red-violet – are so-called **warm colours**. In colour perspective (→ p. 19) the cold-warm contrast supports the three-dimensional effect.

    Quality contrast (fig. 67) relates to **chromatic quality**, for example, the degree of achromaticity in the colours – in other words to differences in the colour quality. Küppers also calls this characteristic of aesthetic distinction the **chromatic degree** or **achromatic degree**. An example of this is the contrast between saturated, glowing colours and dull, gloomy and refracted colours.

    In **simultaneous contrast** the colours change their character according to their surroundings. Thus, for example, one and the same colour can seem darker on a light background and conversely lighter on a dark background (fig. 68). Or a so-called warm background makes a hue seem cooler, while a cooler background makes the colours seem warmer (fig. 69).

    Complementary contrast occurs when two complementary colours – for example orange and blue – meet (fig. 70). Complementary colours are opposite each other in the spectrum. When mixed, two complementary colours produce an achromatic colour (→ pp. 12, 13). If two colours are complementary they enhance each other's luminous quality.

    Quantity contrast is based on juxtaposing colour areas of different sizes, e.g. when using large areas of colour and small elements of colour. According to Johann Wolfgang von Goethe, the following quantity ratios make an essentially harmonious impression: yellow-violet: 1:3, orange-blue: 1:2, red-green: 1:1 and red-yellow-blue: 6:3:8 (figs. 71–74).

<u>Examples of contrast in layout</u>

Formal contrasts
– Upright typeface in contrast to italic typeface.
– Sans-serif versus serif typeface.

Thickness contrasts
– Bold type in contrast with normal type.
– White paper surface versus grey effect of print space.
– Light image versus dark image.

Size contrasts
– Large point size versus small point size.
– Large illustration versus small illustration.
– Wide column versus narrow column.

Colour contrasts
– Black basic text versus text accentuated in red.
– Colour versus complementary colour.
– White paper surface versus grey effect of the text.

Area contrasts
– Short page area versus long page area (format).
– Large text blocks versus small text blocks.
– Broad margin versus narrow type area.

Order contrasts
– Horizontal arrangement versus vertical arrangement.

Rhythm, symmetry and asymmetry

Symmetry (ancient Greek: symmetria = uniformity) is the mutual correspondence of parts of a whole in terms of size, shape, colour or arrangement. Objects or elements can be mirror-symmetrical (reflected along the axis of symmetry) or also centre-symmetrical.

75

    Rhythm is created by uniformly structured movement or by a chronological sequence of patterns or events (repetitions). Structures or patterns can be rhythmical, for example (fig. 75). In addition, auditory patterns can be found in music and motoric patterns in dance.

Checklist: Design resources

– Formats and areas
– Spaces and proportions in an area
– Order and arrangement
– Colours, forms and contrasts
– Symmetry or asymmetry
– Structure, rhythm and movement
– Perspective

Laws of design

A **whole** is made up of parts. Here, parts are those components of a whole that are bound up with the structural principle of the whole. The composition, the arrangement of the individual parts and their ordered whole (cohesion) is known as **structure** (fig. 76).

The **laws of Gestalt**, which emerged from studies of **Gestalt psychology**,[14] explain which structures and forms are experienced as Gestalt, in what way and for what reason. As soon as several forms and elements appear together in an area, the eye relates them to each other.

"Gestalt theory is based on the assumption that the perceptual process cannot be fully understood if it is broken down into ever smaller part processes. Perception is more than the sum of these parts – following the maxim: a Gestalt is more than the sum of the individual parts,"[15] says Professor of Psychology Philip Zimbardo. Consequently, only a small glimpse of Gestalt theory can be provided here, but key is that no object is perceived in itself alone or in isolation. The laws of Gestalt today form the basis for every design.

Law of simplicity – also the law of good design form. The perceived elements or objects are always perceived as simply as possible.

Tip: Simple in this case means short, clear and accurate representation of content.

The following design rules are subject to the law of simplicity.

Law of proximity. If grouped elements are placed together, the sections that are the shortest distance apart from each other are perceived as belonging together. In fig. 77, three points appear separate from the others. These form an independent unit (group), as they are a greater distance apart from the other points. In fig. 78, proximity is dominant over similarity of form. The proximity law also applies to texts. Texts are dissected into words by the space between words; the individual letters form a unit as a result of their proximity, or connections in terms of meaning are structured by blank lines within the sections of text.

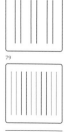

[14] Established by Max Wertheimer, Wolfgang Köhler, Kurt Koffka
[15] Zimbardo, Philip; Gerrig, Richard: Psychology and Life (2001)

Law of similarity – also of equality. Objects or elements that are similar to, or the same as, each other, for example those with the same form or colour, are perceived as one unit by the eye (→ fig. 79).

Tip: Similarity supports the demand for a uniform design principle. Colour, form, size and surroundings are used as ordering devices. However, the more similar, for example, letters, are to each other, the less well they can be distinguished from each other, and the words as images are less easy to read (→ ch. Typography, p. 62).

Law of closure. Parts that are not present are added perceptually; incomplete figures are seen as coherent (fig. 80). Elements that form a complete figure seem coherent and are perceived as a unit (fig. 81). If the laws of proximity and closure clash, the latter is the stronger.

   Law of common fate. If the same elements **share a common fate**, in other words move in a common direction or change uniformly, they are perceived as a unit (fig. 82). Thus, for example, moving objects within a static environment are perceived as a unit.

   Law of continuation. Elements that are **continuously linked** spatially or chronologically appear to belong together and are perceived as a unit by the eye. Here perception is based on experience, and an apparent continuation is accepted.

   Law of symmetry. Symmetrical elements are seen as belonging together, in contrast with asymmetrical elements. Symmetry around the vertical central axis has a more powerful effect than around any other axis.

   The more design laws are simultaneously applied, the clearer their functions become. Fig. 83 shows the proximity law, fig. 84 the proximity and symmetry laws.

Additional law

Law of experience. The only difference between the two objects (fig. 85 and fig. 86) is that one (fig. 85) is upside down in relation to the other (fig. 86). Goethe sums up the meaning of this law in the following words: "One knows only what one sees, and one sees only what one knows".

Figure-ground distinction

Figure-ground. In design, the object that is simpler in form becomes the figure, while the object with the more complex form becomes the ground (figs. 87, 88). The distinction as to whether something is figure or ground (and thus neglected), is not dependent on the colour but on the **distinctiveness** of the figures.

In fig. 89, size is the key factor in the figure-ground relationship: the black lines form the figure, while in fig. 90 there is no figure-ground distinction.

But a changing perception of the figure-ground relationship can also be appealing: So-called **ambiguous patterns** are drawn structures that are deliberately designed so that optical perception admits two different interpretations (**ambivalence**). Victor Vasarely[16] called this the "kinetic effect". A familiar example of this is the **Necker cube**[17] (fig. 91). The drawing is a wire-frame model of a cube. But the cube looks different according to how we focus on it: it looks like either a cube beginning bottom left that we are looking at from top right, or a cube at the top right we are looking at from bottom left. The two overlapping squares can thus be seen as either the front or the back.

Other familiar examples are the ambiguous figure of Edgar J. Rubin[18] (fig. 92) and the Thiéry figure (fig. 93). The works of Maurits Cornelis Escher (→ p. 49, Tips and Links, Perspective and illusions) are also worth mentioning. Ambiguous figures are related to puzzle pictures, in which a certain object has to be found somewhere in the picture.

[16] French painter and graphic artist of Hungarian origin (1906–1997)
[17] Named after the Swiss geologist Luis Albert Necker
[18] Danish psychologist (1886–1951)

Signs and Symbols

Semiotics – also sign theory – is the theory of signs, sign systems and sign processes. It forms the basis of communication theory.

In linguistics, a semiotic process takes place when a message is dispatched from a sender to a recipient and the recipient can decode the message (fig. 94). The recipient classifies and interprets the information. This process puts the recipient in a position to interact with the sender. The sign, marked in a particular code, e.g. **language**, describes its object and is interpreted by the recipient.

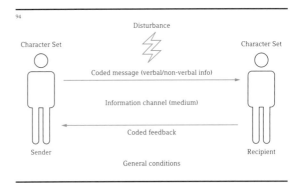

94

Disturbance

Character Set   Character Set

Coded message (verbal/non-verbal info)

Information channel (medium)

Coded feedback

Sender   Recipient

General conditions

Semiotics studies syntax, semantics and pragmatics. **Syntax**, as a theory of connections between the individually marked signs, addresses formal aspects (form) and their interconnection. **Pragmatics** deals with the relationship between senders and recipients, the sign users. **Semantics** looks at the meaning of something marked. It refers to something and thus expects it to be interpreted.

The sign

According to the Swiss linguist **Ferdinand de Saussure**,[19]
a sign is the connection between something **signified** and
a **signifier**. Here, whatever is signified, also called signifié,
represents the conceptual content side of the linguistic sign,
and a signifier, also called signifiant, represents the expressive side of the linguistic sign.

For **Charles S. Peirce**, the founder of modern semiotics,
a sign is something that everybody perceives as standing for
something in a certain respect.[20] Umberto Eco, the Italian
art philosopher and writer, recommends calling everything a
sign that can be grasped as something that stands for something else on the basis of a previously agreed social convention. Here he is largely adopting the definition of Charles
W. Morris.[21]

Charles S. Peirce divides signs into three sign classes,
with nine sub-sign classes and ten main sign classes. The
sub-sign classes discuss **icon**, **index** and **symbol**. These
belong to the second category of signs, in which the object-
relation of the sign is addressed.

Types of signs

Signs are perceptible, either visually or audibly (sound signs).
Facial expressions, gestures, series of letters, images, light
and brands can be interpreted as signs, but so can clothing,
architecture or music. Signs can be conveyed in verbal language (orally and in writing), by touch (tactile signs) and by
smell (olfactory signs). Paraverbal symbols, audible components of speech without content, such as vocal tone and
pitch, can also be interpreted as signs.

Signs can be represented visually in various ways:
realistically or schematically, pictorially-illustratively, in a
painterly-artistic way, symbolically, scientifically, as marks
(signatures), abstractly or as signals (traffic signs).

Signs can be formed by design basics such as form,
colour, size, arrangement, texture, spatial quality, movement
and direction.

[19] de Saussure, Ferdinand: Course in General Linguistics (1998)
[20] American philosopher and physicist (1839–1914)
[21] American philosopher (1901–1979), helped prepare the way for modern semiotics

In the broadest sense, **hieroglyphs** are written signs with pictorial character. More narrowly, the term generally refers to ancient Egyptian **hieroglyphic script**. The first word signs existed in ancient Egypt as early as 3000 BC. This is when merely **pictorial script** developed into hieroglyphic script, where the pictorial signs stand not just for the things represented but for words that sound the same (**ideogram** or **concept sign**).

An **ideogram** is a sign that represents an entire concept, and thus uses symbolic signs for abstract terms.

The **symbol**, originally used as a synonym for the word "sign", represents a thing, an abstract idea or a process. Often the meaning of symbols cannot be rationally translated or interpreted. Two things that have nothing at all to do with each other at a first glance are connected by association, like for example rose and love. The symbol is a sign based on agreement. **Physical**, **chemical** and **mathematical signs** are also symbols.

In psychological terms, a **symbol** has a deeper meaning. The meaning of a traffic sign, for example, is fixed or precisely defined, but the significance of religious or mythological symbols goes beyond the rational plane and has an additional psychological meaning that often cannot be interpreted unambiguously. The clefs used in music are also symbols (fig. 95: treble clef, fig. 96: bass clef).

A **pictogram** – also picture sign – is a pictorial symbol with a fixed meaning (→ ch. Typography, p. 63). Pictograms are modern signposts. Their job is to pass on subject matter and information unambiguously, quickly and very strikingly. A pictogram performs many functions: it can refer, indicate, forbid, warn, report conditions, designate, name or classify (fig. 97).

Marks or **seals of quality** are used to guarantee the quality of goods or services. **Emblems** are signs indicating belonging to a group, a state or a family.

Signet (Latin: Signum = sign) is today used as another word for logo.

The logo

The **logo** is a graphical or typographical sign representing a company or an institution and is used to distinguish between the same goods and services from different manufacturers or providers. The logo, the **mark**, is part of a company's visual image, which is also known as **corporate design** (→ ch. Marketing, p. 238). A logo can consist of one or several letters and be a word mark. It can also be an image or a combination of the images and letters, the word and design mark (→ ch. Law, p. 272). Fundamentally, a logo should be unmistakable, succinct and memorable.

Adrian Frutiger says on this subject: "The secret of beautiful form lies in its simplicity. The best logo is the one that a child can copy in the sand with one finger."[22]

It must also be possible to reproduce a professional logo. A logo should meet the following **design criteria**: attitude, presence, universality, succinctness, recognisability, consistency and economic viability.

Note: A recipient's ability to absorb information is limited; overstimulation reduces this further.

Trends and development

The general situation for designers has changed radically in recent years. The world of brands and trademarks is undergoing a structural change; at the same time, social and commercial developments mean that brand conventions need to be reconsidered and new approaches taken. This change is driven by globalisation, virtualisation, the constantly increasing range of products available and the ultra-rapid development of new markets world-wide. Workers in creative fields and their clients are thus faced with new challenges. The challenge in future will be to link approaches that have stood the test of time with new ones.

---

[22] Frutiger, Adrian: Eine Typografie (1995). Not available in English.

These phenomena manifest themselves in different ways, and it is possible to discern opposing tendencies in terms of globalisation and the mass market. Consequently, a logo has to assert difference more forcefully than ever, yet also demonstrate its belonging to a particular community. Concepts such as "**fractalisation**" and "**social marketing**" become crucial.

It is easy for design freedom to be taken to excess in the world of logos: the structure of logos is evolving into something flexible, variable, dynamic and complex. The pictorial language is becoming more tangible and direct, the messages more forceful and emotional. Immaterial symbolic values, unforgettable experiences and also individualism are all part of the so-called new communication. Today, seasonal campaign logos are produced, for example. Powerful "global brands" are adapted and applied to local living conditions or homogeneous spaces. Free, overcharged ornaments and collages are created, rather than austere forms. Brands are presented theatrically, and whisk us off to magic worlds.

When **designing a logo**, form, colour, proportions and lettering can be used as creative elements. As a rule, the form of a logo is equal to its material limitations. Various basic elements (→ pp. 25–27) can be used for the form. Choosing the right typeface is a key factor here. It gives the logo character, and thus becomes the logo itself.

The following questions should be considered when designing a logo:

- What overall impression of a particular brand is the logo intended to convey?
- What aims is the logo pursuing and what statement (image positioning) is it intended to convey?
- What target group is it addressing?
  Note: For a logo's intended statement to be clearly comprehensible, it is important to establish the cultural context in which it is to be used.
- Are the colours, typeface and form appropriate to the company's image?

- Does the logo look entirely different from those of its competitors?
- Is the logo clearly legible, comprehensible and memorable; can it be retained and recognised?
- Will the logo transfer effectively to all media, and can it be reproduced using all the usual printing processes?
- Can the logo be used in every conceivable advertising medium?

Tip: *The logo should be available in several variants and different file formats for presentation to the client. Highly specialised printing techniques should be avoided.*

- Is it still fully effective and impactful if reproduced in monochrome?
- Can the logo be used in positive and negative form?
- Can the logo be enlarged or reduced in size without loss of quality (legibility and impact)?
  Note: The basic version cannot simply be enlarged or reduced as a whole as proportions can alter. It is better to create and optimise several sizes.
- Can the logo be used in the long term, independently of current trends?
- Does the logo reproduce appropriately in other media, such as fax or on a screen?
- Can the logo colours be transferred to the overall corporate design?
- How economical is the logo?

Tip: *More than three colours or several special colours raise production costs.*

- Should the logo be supported by a slogan (strapline) to help convey the core statement?
- Can the logo be varied for other business areas if required?

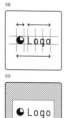

Constructing a logo

The construction of a logo also follows the principle of creative economy and should be developed to produce an image that will last. Thus every part of the corporate design is based on characteristics that remain the same and recur. As a rule, every sign must be disciplined with respect to proportion – and this should be done using a **grid**. All the elements of a logo and their proportions in relation to each other (e.g. proportions relating to size, width and length, distances apart, positions) must be defined precisely.

The **basic framework** – also **module** – is usually a square that is broken down into square modules in its turn. Any additions outside the square form must correspond with the internal modules (fig. 98).

Once a logo has been created, it needs a defined **protection zone** (fig. 99). The protection zone guarantees that no additional design elements (e.g. images, lines or areas of colour) are placed in the logo or the area around it. As a rule, the protection zone is marked digitally, with the aid of a clearly defined white zone in the **CI manual** (→ ch. Marketing, p. 237).

Layout and Page Design

The **draft** or **layout** conveys an impression about the design and the nature of the subsequent final version to the designer and to everyone else involved in the creative process (including the client). It thus serves as a binding basis for decisions about the way forward.

Here, a clear preliminary **briefing** (→ ch. Marketing, p. 219) lays the foundation for successful design. Layout design can be divided into different phases, as described below.

The **sketch** contains the **conceptual idea** as a rough, often freehand, drawing, and conveys the general design effect.

Tip: *Every design should be based on sketches and scribbles.*

The **rough layout** is developed from the sketches. The format is already that of the final format; all the design elements are sketched in more detail so that the overall effect can be judged. The **fair copy** or **final draft** is the last step and is the preliminary stage for pre-press. Then the design is fully worked through and executed in detail.

A **dummy** is a printed specimen that is realistic or at least close to realistic, a **presentation sample**. Make-up is the name for the technical planning and composition of a page according to certain requirements.

Layout process

Work on the layout should not be started until the content is fixed and fully planned, the texts written and the pictures chosen. If pictorial material is not available, it can be ordered from picture agencies (→ Tips and Links, p. 49), and licences acquired (→ ch. Law, p. 268). If a document is completely new, then the following steps make sense as listed:

1. Fixing the page format.
2. Creating the document.
3. Defining the type area.
4. Fixing the design grid with text and stylistic elements.
5. Make-up and final corrections.

3 : 4

2 : 3

5 : 8

1 : √2

1 : √3

1 : √5

A   B

A : B = B : (A+B)

The Golden Section and harmonious page formats

Formats other than the DIN series of **standard formats**
(→ ch. Production, pp. 152–155, Paper formats) are more
suitable for some designs. According to Jan Tschichold,[23] the
page proportions shown here (figs. 100–106) are particularly
well balanced (**harmonious page formats**). The format ratio
5:8 (fig. 102) corresponds to the so-called Golden Section;
the number ratios of the **Golden Section** (→ ch. Typography,
p. 82) tend to be simplified in practice. The Golden Section is
expressed as the ratio 1:1.618. It is not only the epitome of
ideal proportion in art and architecture, but can also be
shown to occur in nature.

The **Golden Section** is based on dividing a line in two
unequal parts (fig. 106), of which the smaller part (A) relates
to the larger part (B) as the larger part (B) relates to the sum of
both parts, i.e. the entire line.

If the format is not explicitly fixed in the briefing (→ ch.
Marketing, p. 219), it is worth experimenting with page
formats. But beware: formats with pages of almost the same
length do not generate any excitement.

The mathematical basis of the Golden Section is the
Fibonacci series 0, 1, 1, 2, 3, 5, 8, 13, 21, 34, 55, 89, 144,
233, …[24] The next number in this sequence derives from the
sum of the two previous ones.

The design grid

The design grid, also construction grid, helps to organise a
design area. It makes it easier to design clearly, consistently
and with an eye to continuity. A design grid is there to divide
the area being designed in such a way that the type area
(→ p. 43) consists of a grid of fields (grid fields), in which all
the stylistic elements concerned such as text or illustrations
can be arranged (fig. 107). A design grid can be constructed
on the basis of the Fibonacci series.

[23] Jan Tschichold (1902–1974) was a noted German calligrapher, typographer, author and teacher.
[24] Named after Leonardus filius Bonacij, known as Fibonacci (13th century), master arithmetician and the
most important medieval mathematician.

Checklist: Design grid

The following design elements should be established before
one begins with the layout:

1. Format.
2. Type area and margins.
3. Columns (column width, column spacing, column height).
4. Basic typeface and point sizes (number of characters in
   a column).
5. Column spacing (number of lines in the column).
6. Dividing the column into grid fields.
7. Margin types.
8. Pagination (→ ch. Typography, p. 82).
9. Colour climate.

Long lines of text are tiring to the eye and detract from ease
of reading. To avoid unduly long lines of text (→ ch. Typography, p. 76, Setting alignment), the type area is divided
into columns (fig. 108). The type area can contain one, two,
three, four or more columns. Here the basic typeface size
(→ ch. Typography, pp. 60, 61, Perception and reading behaviour) and set width have an important part to play.

Tip: *Remember that the more columns a type area has the
less text it can accommodate. The distance between columns
should clearly demarcate the columns of text. It is recommended that the column spacing should be based on the width
of the pair of letters "mi" in the typeface used.*

The **base line grid** can help to set all lines in the **correct
register**; the lines on all the pages are placed to exactly the
same line, also known as **keeping to register**. The width of
the base line grid is based on the line spacing (→ ch. Typography, p. 75) of the basic text. The base line grid begins at
either the upper edge of a page or the type area (fig. 109).
As a rule, only the basic text is set to the base line grid.

Tip: *If the type area contains pictures, the base line grid is shifted upwards by the size of the space between the upper edge of
the type area and the ascenders of the text lines (fig. 110).*

Creativity

Creativity (Latin: creare = to create) is the ability to adapt knowledge and experiences from various areas of life and thought, interpret them in a new way and thus break away from existing patterns of structure and thought. Anyone who wants to be creative must be able to outwit his brain skilfully, says the brain researcher Ernst Pöppel.

Creativity is also a synonym for resourcefulness, ingeniousness, remarkable qualities, innovation, originality, productivity and creative force. According to creativity researcher Joy Paul Guilford, creative personalities stand out for their heightened sensitivity to problems; their thinking is very fluid. As well as this, their minds are highly original and flexible (divergent thinking).

Creative techniques

The aim of creative techniques is to find ideas or approaches to solutions for a given theme. These methods should be as free from constraints as possible; usually this will be group work, promoting synergy effects and free association (lateral thinking). Some such methods are listed below.

An **idea book** or **sketchbook** is kept to note everything that occurs to designers or impresses them in passing. This includes their own drawings, textual drafts or collected quotations, collaged images, photographs or collections of sample materials. In this way, ideas and sketches that have been captured can lead to new thinking or even represent a solution for a new piece of work in their own right.

Brainstorming

Brainstorming is a method developed in 1953 by Alex
F. Osborn in the USA.

In essence, anything goes in brainstorming. But evaluations, comments, corrections or criticism are forbidden,
as everyone participating in such a session should be able
to express his or her views freely and without disturbance.
Some people even recommend allowing for human bio-
rhythms; times between 9 a.m. and 1 p.m. or 4 p.m. and
8 p.m. are seen as particularly advantageous.

A brainstorming session starts by defining the task at
hand and setting a time limit, as a rule 15 to 20 minutes.
Having a **chairperson** or **discussion leader** can be helpful
here. He or she can keep an eye on the **rules**, explain the
theme or problem and control the communication flow.

All the participants must contribute their knowledge and
ideas, even if they do not seem relevant to them. The fact is
that the free association of ideas, images or objects can introduce the group as a whole to new concepts for solving the
problem. Here, too, focusing on the problem must therefore
remain in the foreground, as over-hasty evaluations make it
difficult to come up with alternatives. It can also be an advantage if there is little consensus between the participants,
as different opinions can lead to innovative ideas. Quantity
comes before quality, as the key is to concentrate on producing new ideas. Any attempt at criticism or taking up a position during the session should be avoided or postponed.

Tip: *There is no individual copyright in ideas produced in sessions of this kind. Brainstorming is about picking up ideas and
spinning them further. Thus individuals cannot claim ownership of either the result or any part of it* (→ ch. Law, p. 268,
Protection of Ideas).

Minutes should be kept to **record the results**. One or two
people should be chosen as non-participants in order to do
this. All of the suggestions made should become part of the
minutes. The minute-takers are not allowed to ask questions, as this interrupts the flow of ideas. The suggestions
made are numbered and kept anonymous. The minutes are

then produced either on paper, on a board or on a **flipchart**. The latter can be stimulating, as the participants can relate more easily to opinions and suggestions that have already been expressed. A summary of all the key points is read out shortly before the end of a brainstorming session, to boost the so-called final spurt.

It is only after the brainstorming phase that the ideas and approaches that have been noted are structured and discussed, and any suggestions that cannot be realised are eliminated. These ideas sessions can only provide raw material for possible concepts, completed solutions are hardly to be expected here.

Mind mapping

The **human brain** has a left-hand and a right-hand **hemisphere**. Research has shown that the two halves of the brain perform different functions. They are mutually activated when thought takes place. The left-hand cerebral hemisphere is responsible for sequential thinking in a right-handed person, e.g. for logic, language, number sequences or linear and numerical thinking. The right-hand half is responsible for holistic thinking, e.g. for expressing and acknowledging emotions, geometrical thinking, spatial perception and structural thinking. So the right-hand cerebral hemisphere is also important for musical abilities, and for the ability to recognise complex images.

The concept and method known as **mind mapping** was developed by the Englishman Tony Buzan in 1974. It is based on insights from research on the human brain. He used these to create a work presentation method that addresses the right-hand and left-hand **cerebral hemispheres** equally by combining linguistic and logical thinking with intuitive and pictorial thinking.

Mind mapping is a graphical variant on brainstorming. The typical structure of a **mind map** resembles a tree structure (fig. 111).

The subject, in other words the central question to be addressed, is written in the middle of a sheet of paper (a horizontal format is advantageous). A succinct, slogan-like formulation is important here; it must be appropriate and striking.

The theme is then circumscribed. Mind maps are not based on complete sentences or parts of sentences, but simply on keywords. These should trigger associations and chains of association by linking impressions, feelings and ideas. The **keywords** allocated or subordinated to the central theme in terms of content are written on lines which form the main branches, and which can then branch further for subsequent sub-concepts.

If further variations on these ideas come to mind, an additional branch is added to the appropriate main branch. This then produces further little branches on the existing main branches. The resultant mind map can be re-organised and re-sorted subsequently, as it is not clear at the outset how this map will develop, and in what direction.

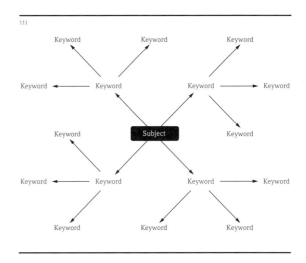

111

Golden Rules of Design

1. Design is a service. Work and design to address clients and target groups.
2. Be creative, look for new solutions and ways of thinking.
3. Find a clear and universal design, stick to visual constants to raise the perception value (identity, recognisability, trust).
4. Keep it short and simple: reduce and simplify to achieve an efficient effect.
5. Design comprehensively, credibly and appropriately.

Tips and Links

Picture agencies

50 international picture agencies on the same website: www.fotosearch.de

Corbis: pro.corbis.com

Getty Images: www.gettyimages.com

Colour

Drew, John; Meyer, Sarah: *Color Management. A Comprehensive Guide for Graphic Designers.* RotoVision, East Sussex (2005)

Free Colour Wheel Calculator: www.sessions.edu/ilu

Homann, Jan-Peter: *Color Management.* Springer Publishing, New York (2005)

Itten, Johannes: *The Art of Color. The Subjective Experience and Objective Rationale of Color.* Wiley, Hoboken (1997)

Perspective and illusions

Ernst, Bruno: *The Magic Mirror of M. C. Escher.* Parkwest Publications, Miami (1987)

Signs

Frutiger, Adrian: *Signs and Symbols: Their Design and Meaning.* Watson-Guptill Publications, New York (1998)

Klanten, Robert et al.: *Los Logos.* Die Gestalten Verlag, Berlin (2002)

Lionni, Pippo: *Facts of life.* Laurence King Publishing, London (2001)

Lionni, Pippo: *Facts of life 2.* Verlag Hermann Schmidt Mainz (2001)

Mutabor: *Lingua Grafica.* Die Gestalten Verlag, Berlin (2001)

Painting and drawing

Smith, Ray: *The Artist's Handbook.* Knopf; 1st American edition (1987)

Perception and psychology

Arnheim, Rudolf: *Art and Visual Perception. A Psychology of the Creative Eye.* University of California Press, Berkeley (2004)

4.0

Throughout human existence as we know it, script has served as a mirror of society. From ancient cuneiform writing hewn in stone right up to modern-day lettering, it not only allows us to visualise, store and communicate language over distance and time, but also reflects the cultural, architectural and technological accomplishments of its age. Up to this day, every century, every decade has left its mark on typography.

Nevertheless, society tends to underestimate the power of letters, often reducing typography to a mundane means of imparting information. This is a prejudice we would like to dispel for good. From Gutenberg's basic, movable letters that revolutionised the printing process to today's digital typesetting that gives rise to almost inexhaustible style variations, this chapter illuminates the historic evolution of various well-known type styles, introduces measurement systems, font technologies and layout rules and translates these for everyday design practice.

Beyond these fundamentals, we also touch on the indispensable peripherals often neglected in the creation and use of typography, e.g. specific monitor typography, punctuation idiosyncrasies, harmonious spacing and the psychological impact of different typefaces.

From unostentatious, functional fonts optimised for clarity and legibility to elaborate visual signs that reflect the zeitgeist of the period in which they were created, this chapter features a broad range of examples. And anyone curious about the revelance of gutters, shoulders, widows and orphans in this context will find the anwers here.

Typography

4.1	**Typography**	56
4.2	**Measurement Systems**	57
4.3	**Script**	60
	Perception and reading behaviour	60
	Origins of type	63
	The letter	67
	The character set	69
	The type family	70
4.4	**Typesetting**	75
	Setting alignment	76
	Accentuation	78
	Writing conventions	79
	Page layout	81
4.5	**Typeface Design**	84
4.6	**Font Technology**	90
	Font formats	90
	Fonts on a display screen	93
4.7	**Tips and Links**	96

Typography

Typography fixes language and makes it visible. The term **typography,**[1] originally a synonym for the **art of the printed book**, now tends to be viewed in a rather more sophisticated way.

Typography creates logical and visual links within an arrangement of letters and other characters appropriate to the task in hand. Typography's principal role is to make a text look as appealing as possible and thus make reading an effective experience – in other words to captivate the reader. Typographic designers create emphases and signals, and navigate the **recipient** (reader) through the text. The designer and typographer Otl Aicher said in this context: "Typography tries to find the right type sizes and quantities to please a demanding eye, and satisfy it." He continues: "In this respect, typography is nothing more than the art of discovering what the eye likes, and offering information so temptingly that the eye cannot resist it."[2]

So one crucial aspect is choosing the shape of the letters, and the way they are organised creatively, fitting them together to create words, lines and paragraphs; this is called **microtypography.** Another important aspect is formatting, shaping and ordering the textual structure in an area, and this is known as **macrotypography.** Even the colour of the paper triggers responses that can change a reader's readiness to accept the material. Typography also includes designing the characters themselves, and realising them technically (type design).

---

[1] Greek: typos = stroke, graphein = to write
[2] Aicher, Otl: Die Ökonomie des Auges (1989). Not available in English.

Measurement Systems

Typography uses its own **measuring systems** and **units**.
In **hot metal setting** (→ pp. 63 – 67, Origins of type), each
individual letter was a so-called **punch** and smaller than the
shank (fig. 1), whose vertical dimensions determined the
type size[3] of a typeface. The (non-printed) space surround-
ing the letter was called the **shoulder**. The body height was
given in points. The **typographical point** is the smallest unit
in a **measurement system** for fixing the size, spacing and
line height of a letter. Various typographical measurement
systems have emerged in the history of typography. Even
though the **body** is only a fictitious value now in electronic
setting systems, type manufacturers still keep to traditional
standard dimensions. The most common typographical
measurement units today are **DTP point sizes** and **Didot
point sizes**. The DTP point size (also PostScript point) is
based on 1/72nd of an **inch**. As 1 inch = 25.4 mm, the DTP
point thus represents about 0.352 mm. The most important
typographical measurement units with their conversion
values are listed in the table below.

1

Unit	Abbreviation	Relative size
Cicero	cc	12 Didot points
		4.5 mm
Didot point	dd	0.375 mm
DTP point	pt	0.352 mm
		1/72 inch
Inch	in	25.4 mm
		72 DTP points
Millimetre	mm	2.85 Pica points
		2.67 Didot points

---

3   Not identical with the optical size of a letter.

	Cap height in mm	De-scender	Film-transp.	
4p	1.00	0.30	1.50	Hng
5p	1.20	0.40	1.75	Hng
6p	1.50	0.45	2.25	Hng
7p	1.75	0.50	2.50	Hng
8p	2.00	0.60	3.00	Hng
9p	2.25	0.65	3.50	Hng
10p	2.50	0.70	3.75	Hng
11p	2.80	0.80	4.25	Hng
12p	3.00	0.90	4.50	Hng
13p	3.25	0.95	5.00	Hng
14p	3.50	1.00	5.25	Hng
15p	3.75	1.10	5.50	Hng
16p	4.00	1.20	6.00	Hng
17p	4.25	1.25	6.50	Hng
18p	4.50	1.30	6.75	Hng
19p	4.75	1.40	7.25	Hng
20p	5.00	1.50	7.50	Hng
22p	5.50	1.60	8.25	Hng
24p	6.00	1.70	9.00	Hng
26p	6.50	1.90	9.75	Hng
28p	7.00	2.00	10.50	Hng
30p	7.50	2.20	11.25	Hng
32p	8.00	2.30	12.00	Hng
34p	8.50	2.50	12.75	Hng
36p	9.00	2.60	13.50	Hng
40p	10.00	2.90	15.00	Hng
44p	11.00	3.10	16.50	Hng
48p	12.00	3.50	18.00	Hng

Line weight of the Linotronic
— 1   0.10 mm
— 2   0.20 mm
— 3   0.40 mm
— 4   0.80 mm

Various methods are available for measuring type size. The commonest measuring units are the typographic point and the millimetre. New media typography also includes the pixel unit (→ p. 93). As most of today's computer programs work with American units, modern type gauges or rulers (→ p. 58) use the DTP point system alongside the Didot point system.

Type sizes

The following factors are crucial for establishing type size: the purpose for which the type is being used, the medium and the stylistic characteristics of the typeface. Type setters divide type sizes into the three following groups:

Reference sizes are type sizes of 5 to 8 point. They are used for side notes (→ p. 82, marginalia) and footnotes as well as in reference works, telephone directories, dictionaries etc.

Text sizes include type sizes of 9 to 12 point. They are used for continuous text, e.g. in books (→ p. 62, text typeface).

Display sizes (also distance sizes) include type sizes from 14 point (fig. 2). They are used for headings (titles) or for text elements intended to be read at long distance, e.g. for posters.

2

Sample word [14]

Sample word [24]

Sample word [36]

Script

A script consists of a repertoire of defined symbols whose meaning (→ ch. Design, pp. 34–36) is based on agreements forged over the centuries. Recognising and identifying signs is essential if the reading process is to function.

Perception and reading behaviour

Within a culture, the **anatomy** of a script, in other words the **inner structure** of a letter, is laid down clearly. Thus, in our Roman script alphabet, the shape of individual letters can be changed only within certain limits, if they are to be easily decoded. Meanings are allotted to the sequences of letters (**semantics**). Here, words are recognised more or less automatically.

The meaning of a visually recorded image is often worked out only slowly and inadequately, while the meaning of a word-image is often decoded precisely and quickly. By analogy with the spoken word, purely rational comprehension of visual images is impossible.

Reading means **conscious perception** and **recognition** of meaning and also subconscious perception of the script that conveys this meaning. When we read, we think that our eyes are moving continuously over the text we are trying to understand, and that they absorb it continuously – like a scanner. But this is a mistaken impression: our eyes either stand still for a moment, fixed on a particular point and its immediate surroundings, or they make short, jumping movements, **saccades**,[4] lasting about 20 to 35 milliseconds.[5] The length of the jumps depends on the difficulty of the text and covers about seven to nine letters (fig. 5).

So we do not take in individual letters when we are reading, but whole words and groups of words. The main information is drawn from the upper part of a script (figs. 3, 4).

3

4

Saccades are the short, jumping movements our eyes make

when they read. Between the jumps, which occur approxi-

mately every seven to nine letters, our eyes fixate on points.

If we have text comprehension problems, our eyes make

jumps backward that are also called regressions.

The **legibility** of a script depends largely on how it is designed formally and aesthetically, and how the script is handled in terms of design and typography. Different demands are made on scripts, as reading habits, ways of reading and the reading situation have to be taken into consideration. In the case of a long, coherent text, also called **body** or **continuous text**, the important factor is that the content or the general context can be grasped rapidly and without difficulty. Even small changes can affect the legibility of a text considerably. But for headings (→ p. 62, headline typeface), advertisements or posters, it is important to attract the readers' attention first of all.

<u>Typeface and expression</u>

Reading matter is absorbed by grasping signs visually. **Visual signals** trigger **associations** and sensations that can be specifically influenced through the external structure of the letters, e.g. the expression of a font (fig. 6), or through the typographic design.

Caution wet paint!

The **level of response** to a typeface – influenced by dimensions, organisation and accentuation – is determined to some extent by the **typographer's** intentions, but above all by the text's purpose and function.

A typeface must be appropriate to content, aims and function. If a **text typeface** (also called text type, basic type, body type) has too vigorous a life of its own it can inhibit reading fluency (→ pp. 84–89, Typeface Design). This effect is sometimes harnessed for effect in typographic design for headline typefaces.

The image of a typeface, its character, is determined by its **characteristic style (ductus[6])**; this style is reflected in its lines, originally determined by the characteristic way the writer handled his pen.

A **headline typeface,** also called display face or title face, can be emphatic, sensational, pictorial, provocative or expressive. Headline typefaces are usually large, used mainly for headings and titles or for emphasis (→ p. 78, Accentuation). A headline face's specific function is to demonstrate presence. The possibilities here range from subtly refined to radically wild. Headline faces can form independent textual units or support the basic text. The crucial factor is the directly perceived outward impression made by the **type or word image** (fig. 7). Thus typography becomes emphatically an **expressive medium.**

7

Legibility decreases

Pictorial quality increases

Text type · Display type

Display faces are outstandingly well suited to translating moods into visual form. If expression or originality are of primary importance, this usually happens at the expense of legibility, but without compromising the lucidity of the statement. Breaking the rules of harmony makes a major impact as well. Often a single set of lettering is used for a headline face or individual letters are developed, and the rest of the alphabet is ignored (→ pp. 84–89, Typeface Design).

[6] Lat.: ductus = leading

If objective, more **reticent typefaces** are required for titles,
it is possible to use bolder weights (→ p. 70) of existing faces
(→ p. 78, Accentuation).

Origins of type

The formal language of type is inextricably linked with
humanity's social development. Every form of type that has
developed historically reflects architecture, technical and
cultural achievements (also writing implements, materials)
that are closely related to the human mind in a particular
era. They give us a sensual and aesthetic insight into different periods.

The first script that justifies the name is **cuneiform**
(approx. 3,000 BC). But the first **representations of numbers** and **counting signs** date from about 30,000 BC. The
versions of **Sumerian cuneiform** that have come down to us
contain laws, contracts and social communications.

Over the millennia, script developed in various stages:
pictures, **pictorial signs**, **pictograms**, **concept signs** (ideo-
grams) and **hieroglyphics** (→ ch. Design, p. 36).

The **first recorded alphabet** (→ p. 68) in the world, from
which our **Latin alphabet** stems, was created in approx.
1,200 BC by the **Phoenicians**. Its development was triggered
by the Egyptian system of hieroglyphs. Fig. 8 shows the
Phoenician symbol for "A" at around this time.

The **Greeks** took over the 22 **Phoenician phonetic
consonants** and added vowels to them. They developed an
alphabet that has shaped many Western writing systems.

The Roman **Capitalis Monumentalis** or the **Capitalis
Quadrata** (fig. 9), as amended for books, still forms the basis
of the capital letters (→ p. 67) we use in our Latin alphabet
today.

But the small or lowercase letters (→ p. 67) in our Latin
alphabet originate from the **Caroline Minuscule** (approx.
AD 800, fig. 10), which became a reliable **script for correspondence** that could be written and read quickly under
Charlemagne. **Gothic Minuscule** (approx. 13th century AD)
was one of the scripts to develop from Caroline Minuscule.
This was a narrow script that could be written quickly, con-

8

9

10

11

sisting of letters that were very close together with broken lines and a grid-like structure (→ fig. 11). Gothic Minuscule is also a forerunner of **German Gothic** script (16th century).

In around 1440, the coin-maker and goldsmith Johannes Gensfleisch zur Laden, known as **Gutenberg**, invented the casting of **movable type**. The process of putting individual letters together to compose and reproduce printable pages is largely the result of his invention. In this way Gutenberg forged the path from written to set script (hot metal setting), which meant that ideas and knowledge could be copied mechanically.

The most impressive single item from this period is the Gutenberg Bible (1452–1454), written in Latin. Letterpress printing developed at a great rate after 1470, and made it easier for information to be disseminated very quickly to members of society who had hitherto had little access to it. Gutenberg remained firmly committed to the Gothic principle with the **Textura** font he used for printing his Bible (approx. 1456). At the same time, the so-called **Roman** typeface developed as a new font in Venice; it still provides the basis for our printed Latin alphabet.

Humanists in the **Renaissance** (from the 15th century) chose to go back to the ancient Greeks and Romans for their models, and thus also affected the form of the Roman typeface, which brought upper and lowercase letters together for the first time. The Roman typeface went back to ancient forms: **Roman capitals** for the **uppercase letters** and Caroline Minuscule (humanist) for the **lowercase letters**. The **formal canon** of our present-day Latin alphabet is thus based on two fundamentally different alphabets. **Arabic numerals** were also added to our present alphabet during the Renaissance period. These **numerals**, now international, replaced Roman numerals when the decimal system was introduced.

One of the most impressive Roman typefaces from the **Old face** (also Geraldics) is still **Garamond** by the type-cutter **Claude Garamond**. The characteristics of this type are the small degree of contrast between **hairline** and **main stroke** (→ p. 68, fig. 23), the soft line (→ p. 62, ductus) and the left-leaning axis (→ p. 74, fig. 2) of the rounded forms.

Transitionals (mid-17th century) show greater contrasts between hairlines and main strokes, the axes of the rounded letters are less diagonal, and the serifs (→ p. 74) finer (e.g. the "Janson text" typeface). They bridge the gap between **Old** and **Modern typefaces**. Modern typefaces (also Didone) clearly show the influence of **copperplate engraving**, which made the extremely marked contrast between hairline and main stroke possible. New forms were constructed; the type axis (→ p. 74) stands vertical, the handwriting line disappears. Those most involved in developing these typefaces (17th–18th centuries) were the **Didot** family of printers (→ p. 57) and the Italian **Giambattista Bodoni**, as well as the German Erich Wahlbaum in the 19th century. The tendency towards standardisation in this period also produced new standards for typography (→ p. 57–59, Measurement Systems).

Industrialization in the early 19th century brought about the **slab-serif faces** (e.g. **Egyptian faces**) (→ p. 73, fig. 5) in England. These typefaces contain no or very small differences between the hairlines and main strokes as well as emphasised or monumental-looking serifs.

The first **sans-serif** typefaces were also developed in England at this time. The technical appearance and the even-looking line weight without serifs (fig. 12: the simplified form of "g") were called "**grotesque**" by contemporaries, and the name stuck. The use of these new typefaces (e.g. **Akzidenz Grotesk**) was initially intended exclusively for setting titles and headlines (→ p. 62) in posters and advertisements. As a new reproduction technique, it was in particular lithography, that made it possible to liberate typefaces. In lithography hand-drawn fonts and free letter forms could now be reproduced in terms of printing technology.

12

The **20th century** transformed the **medium of type**. In the early years, different styles, tendencies, **presentation** and **reproduction techniques** took up rival positions and redefined the way type was treated. So-called visionary, radically aesthetic artistic movements and other counter-movements (**Dadaism**, **Constructivism** etc.) on the one hand, and typographic design aimed exclusively at **applied art** on the other attempted to hold their own against classical typography.

13

14

15

16

17

At this time, type increasingly left the realm of the printed word. So, for example, the **Elemental Typography** of the 1920s experimented with formal clarity of expression and geometrically constructed **sans-serif forms** without flourishes such as **Futura** (fig. 13) or **Gill** (fig. 14).

In the mid-20th century sans-serif faces enjoyed a revival. They became the basis for new "modulated" alphabets that deviated from the originally constructed formal principle. **Adrian Frutiger** published his **Univers** (fig. 15) and **Max Miedinger** cut his **Helvetica** (fig. 17) on the basis of **Akzidenz Grotesk** (fig. 16).

A new **era** of typography began in the 1970s. The **opto-mechanical typesetting system (photosetting)** was introduced, and within a decade, hot metal setting, which had been used for 500 years, had all but disappeared. In the late 20th century the mass media produced a wide variety of universal typefaces and approaches to type (→ p. 72, hybrid typefaces).

Technical innovations such as desktop publishing (**DTP**) and the **World Wide Web** revolutionised the type medium and meant that conventions were re-thought, and new paths embarked on.

Type left its traditional vehicles and conquered virtual space. Types were no longer just **analogue**, but now above all **digital** (→ pp. 90–95, font technology, and also ch. Digital Media, pp. 134, 140). Type culture experienced a radical structural change such as had not taken place before, and that was to have a more or less lasting influence on our **reading and seeing habits** (→ pp. 60, 61, Perception and reading behaviour) in subsequent years.

Because it could be reproduced at will and had to an extent been dematerialised, it seemed as if everything was possible in the medium of type – and everything was tried out. Design found a new freedom; typography and script were **democratised**. Hardware and software were available everywhere (→ ch. Digital Media, pp. 111–113), which enabled pretty well anybody to work with a typeface of their own (→ pp. 84–89, Typeface Design), to generate it or to modify existing material. Typefaces were suddenly popular.

Recourse to history meant new typographical adaptations and initiatives; epoch-making type styles were collected like objets trouvés and reassessed. **Archaic motifs** were illustrated in a typographically appropriate way, or anticipated. Typeface, typographical technology (→ pp. 90–95) and the type market clearly showed inflationary traits, but this also triggered new challenges, approaches and technologies (e.g. PostScript Type 1; → pp. 90–95, Font Technology). Screen- and pixel-adapted type forms (fig. 18) were created for reading on screen (→ pp. 93–95, Fonts on a display screen).

    For economic reasons and because of the aesthetics of the day, typefaces for continuous text (→ p. 61) became increasingly narrow and their x-heights (→ p. 68, fig. 23) greater; their forms also became more open.

The letter

The letter is the graphic shape of a grapheme or character used to represent linguistic sounds. In modern desktop publishing, letters are also called glyphs. In typography, a glyph is a graphic depiction of a type character.

    The letter is the smallest element in forming a word or shaping a text; their arrangement is a key factor in determining the typographic form and legibility of a text.

Uppercase and lowercase

Capital letters are described as **uppercase** or more rarely **majuscules**. Small letters are known as **lowercase** or more rarely **minuscules** (→ pp. 63–67, Origins of type).

    Uppercase characters are based on three basic geometrical shapes: triangle, circle and square (→ ch. Design, pp. 26, 27). Here, the forms and directional contrasts are fundamental to the appearance of the letters (→ figs. 19–21). Fig. 22 shows the structural principle for a Renaissance upper case "M" according to Albrecht Dürer (1471–1528).

Hint: See also Typeface Design, pp. 84–89.

<u>Letter anatomy</u>

The sum of all of a letter's elements conveys the essential qualities of a typeface. The individual letter elements are defined using specific technical terms:

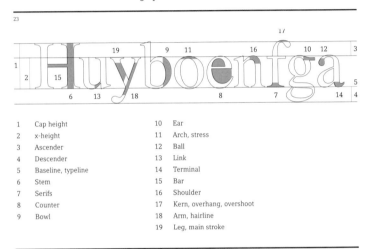

1	Cap height	10	Ear
2	x-height	11	Arch, stress
3	Ascender	12	Ball
4	Descender	13	Link
5	Baseline, typeline	14	Terminal
6	Stem	15	Bar
7	Serifs	16	Shoulder
8	Counter	17	Kern, overhang, overshoot
9	Bowl	18	Arm, hairline
		19	Leg, main stroke

In addition to the visible elements, the areas of white between the letters also play a crucial part. The **interior of a letter** is known as the **counter**.

The **alphabet** is the visual version of all the sound units of a language in a fixed sequence.

The character set

The **character set**, or character stock, of a typeface consists "of various types of characters such as letters (sound
characters), figures (quantity characters) and an imprecisely
defined number of characters for controlling the writing (for
example full stops, brackets and inverted commas)."[7]

Punctuation and special characters

Punctuation and **special characters** are used to impose
structure, order, and value. Punctuation marks reflect the
voice's ability to modulate. They can reinforce meaning, and
draw structure and thoughts, figures and values together.

Arabic numerals

The individual character for expressing **quantity** is known as
a **numeral** (or numeric character). A **number** is produced by
combining different numerals (→ pp. 63–67, origins of type).
In typography a distinction is also made between **lower** and
uppercase figures, which include numerals (→ p. 81, table
numerals).

Ligatures

A combination of several letters fused together is called a
ligature. One example of this is the **ampersand** character or
commercial and. Jan Tschichold[8] says of it: "… written 'et',
comes from Latin, means 'and' and is a ligature of a very
particular kind because it appears in so many forms. It is
always an intimate fusion of letters, in which one part of a
letter either merges into part of another or forms that part
at the same time."

Flusser, Vilém: Die Schrift. Hat Schreiben Zukunft? (2002). Not available in English.

Tschichold, Jan: The Ampersand. Its Origin and Development (1957)

The type family

A **type family** is the full set of types available for a typeface. The **typeface**, the term for a **full set of characters** includes all the characters within a type variant (e.g. Helvetica New 23 Light Extended). It originally referred to a set of characters cut in metal.

	Terms
Type widths	Condensed
	Regular
	Expanded
Type weights	Ultralight
	Thin
	Light
	Roman
	Book
	Regular
	Semibold
	Bold
	Heavy
	Black
Type lie	Regular
	Italic

Hint: See also pp. 75–77, Typesetting.

Lowercase	abcdefghijklmnopqrstuvwxyzæœ
Uppercase	ABCDEFGHIJKLMNOPQRSTUVWXYZ
Accent marks	À È Ì Ò Ù à è ì ò ù Á É Í Ó Ú á é í ó ú
	Â Ê Î Ô Û â ê î ô û Ë Ï ë ï Ã Ñ Õ ã ñ õ
	Å å Ç ç
Umlauts	Ä Ö Ü ä ö ü
Ligatures	Æ Œ æ œ ff fi fl ffi ffl
Punctuation and special symbols	. : , ; ? ! - – —
Parentheses and brackets	( ) [ ] { }
Symbols for currency	$ £ ¥ €
Symbols for per cent, one tenth of one per cent and degree	% ‰ °
Ampersand and asterisk	& *
Symbols for at, registered, copyright and trademark	@ ® © ™
Apostrophe	'
Opening and closing inverted commas (quotation marks)	„ " " " ' ' , ' « » » « ‹ › › ‹
Lowercase numerals	1234567890
Uppercase numerals	1234567890

Type systems are extended type families, also known as
hybrid systems. These type families are not just distiguished
in terms of lie, width or weight, but also in terms of the type
appearance. So different families (with and without serifs)
are brought together as a wider family with the same system
and proportions.

The following overview shows the different stylistic characteristics that a letter in the type system Compatil can have
while maintaining the same proportions (fig. 24).

24

n n n n

Typeface classification

Various attempts have been made to sum up the whole
variety of typeface types in categories; some classification
models have emerged, but they do not yet provide a complete overview.

The DIN standard 16 518 of the German Institute for
Standardization lays down the standard classification system for typefaces in Germany, and takes a chronological
approach. But this standard applies only to hot metal fonts
that were created until the 1970s. This standard follows the
proposals of the Association Typographique Internationale
(ATypI).

<u>Typeface classification according to Vox (1954)</u>[9]

1. Humanes
2. Geraldes
3. Réales
4. Didones
5. Mécanes
6. Linéales
7. Incises
8. Manuaires
9. Scriptes

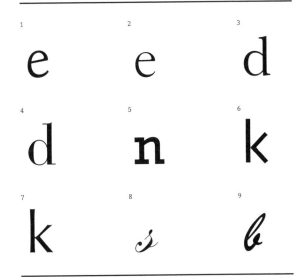

Additional type classification systems include the British Standards for Type Classification (1967) by the British Standard Institution and the PANOSE System (1985) by Benjamin Bauermeister.

[9] Maximilien Vox (1894–1974) was a French typographer. This classification method was adopted by the Association Typographique Internationale (ATypI) in 1962.

Typeface classification according to DIN 16518

1. Venetian Old Style
2. Old Face
3. Transitional
4. Modern Face
5. Slab-Serif
6. Sans-Serif
7. Decorative and Display
8. Scripts
9. Handwritten
10. Black Letters and Broken
11. Foreign Scripts

Typesetting

To convey meaning comprehensibly, far more is needed than the individual letters. It is only the logical use of space between the characters, the way the words are delineated from each other and the arrangement of sentence components that makes content intelligible.

25

The **character spacing** derives from the **width** of a type character. The width applies to the **width of the actual character** (fig. 25) including the **spaces before and after** it (fig. 26) also called sidebearings, i.e. the white areas that are needed for setting or for fitting the words together in a text. The shoulder has determined the distance between the characters since the days of hot metal setting (→ p. 57).

26

Type characters have various **width values**, also character widths. The **set width** is determined by the width of the character (→ p. 86, spacing a typeface) and the distance between the letters (→ p. 84–89, Typeface Design).

Word space derives from the set width or the distance between the letters. Increasing the distance between words or letters is called **spacing**. Word and letter spacing should be consistent in continuous text (→ p. 84–89, Typeface Design).

Line spacing

Line spacing (fig. 27) is the distance between the individual lines of type. It is measured from line of type to line of type. In the days of hot metal printing, the letters were placed on a body (→ p. 57–59, Measuring systems). The antiquated term **leading** goes back to hot metal printing when "leads" or "gutters" were placed between the set lines as "blind material" (additional space), thus determining the size of the leading.[10]

27

Tip: To achieve a balanced setting in continuous text, the white space between the lines should be about one-and-a-half times the x-height (→ p. 68, fig. 23) of the typeface used, a little less for headings.

28

29

Tip: For reading sizes (about 8 to 12 point) (→ p.59) the follow-ing rule is considered to produce a legible distance between lines: type size + 1.5 to 2.5 point.

Solid setting is the term used when the distance between the lines is identical with the type size (fig. 28: type size is 12 pt, distance between lines is 12 pt). If the distance between the lines is greater than the type size the term **leaded setting** is used (fig. 29: type size is 12 pt, distance between lines is 16 pt).

Tip: In principle, an uppercase setting needs more space between lines than a normal setting. Narrow columns need less and broad columns more visible distance between lines.

Setting alignment

30

31

32

Setting alignment is the horizontal positioning of a text with-in a restricted space. The length of a line is also a crucial factor for the legibility of a text. Here, the type size and width of the set material relate to each other directly. The eye finds unduly long lines tiring, and if they are too short they seem too restless and stressful.

Tip: The ideal width for set material in German is 60 to 75 characters per line. To avoid large holes in justified setting there should never be fewer than 40 characters. If justified setting is forced, it is not just the distances between the words but the distances between the characters that are reduced or increased. Hence this electronic command should be avoided from a typographical point of view. And the distance between characters should always be set at '0' as well.

Setting types

33

The overall visual impression made by a text is also deter-mined to a large extent by the alignment of the setting. A set piece of text can be aligned in various ways: **flush left** (fig. 30), **flush right** (fig. 31), **centred** and in **justified** form (fig. 33). Centred type is also known to use a **symmetrical** or **axial setting** (fig. 32). There are also further subvariants of these settings in typography such as flush left ragged setting, flush right ragged setting or the interlaced justified setting.

In **justified setting** all the lines are the same width. This means that there will inevitably be different spacings between words. If justified setting is poorly executed, the gaps between the words (→ p. 81, empty space) can become disturbing "holes". In **ragged setting** the distances between words are the same for all lines. This creates a calm internal structure. Another characteristic of ragged setting is that the lines differ in length. There are various types of ragged setting.

Tip: *Good ragged setting is a great challenge. The ragged area should be worked on manually after the automatic layout phase to avoid word-breaks that can distort meaning and lead to a lack of rhythm in the way the lines fall caused by the electronic setting program (bulges, steps etc.).*

The **grey value** is created by optical inertia: only individual characters appear to be black. The white of the gaps mixes with the black area of the letter to make grey in ideal cases. This happens within the structure of letter, word and line.

Tip: *Squint your eyes!*

<u>Accentuation</u>

Language emphasises a word by means of acoustic signals such as high or low vocal pitch. But type has to signal emphasis visually. **Accentuation** is emphasis of individual or several words, or whole passages of text. Typography distinguishes between the accentuation methods listed in the following table (fig 34).

34

italic

SMALL CAPS

semi-bold type

bold type

s p a c i n g

CAPITAL LETTERS

larger type size

different typeface

coloured **accentuation**

negative

underlined <u>accentuation</u>

35

36

Italic type is mainly used as a "discreet" accentuating variant on an upright typeface.

Strictly speaking, the expression **italics** means a genuine italic script (fig. 35, top). Its letters come close to cursive writing. So-called **oblique** is simply a computer-generated, slanted version of the straight type cut without the typical character of handwriting (fig. 35, bottom).

Small capitals, often referred to as **small caps**, are uppercase letters (→ p. 67) at the optical height of the x-heights (→ p. 68, fig. 23).

Tip: Care should always be taken that genuine small caps are used (fig. 36, top). As the optical line thickness of electronically reduced uppercase characters (fig. 36, bottom, false small capitals) is not identical with the thickness (visual boldness) of the capital letters, this could otherwise produce a less than attractive appearance.

37

Initials[11] – also called initial letters, illuminated letters or decorated majuscules – are letters that are decorated to enhance their character. The illuminated initial is placed at the beginning of a chapter or paragraph. It will be larger than the basic text and can run through several lines. The initial letter can be placed in various ways, but always on the type line of the main text (fig. 37).

Writing conventions

Writing conventions are based on experience and common practice. Practices differ from language to language. For example, in German-speaking countries a comma is used in a column of figures to indicate the decimal places, e.g. 2.000.000,00, while in English-speaking countries a dot (decimal point) is used to identify decimal places, e.g. 1,000,000.00.

---

[11] Latin: initium = beginning

Omission marks. If a word is not written out in full, the missing part is replaced by three points written without space from the rest of the word. If whole words, large parts of sentences and full sentences are omitted, the omission marks are separated from the rest of the texts with a space on both sides (fig. 38).

Opening and closing inverted commas (also called **quotation marks**) are used differently according to the language concerned.

German-speaking countries	" "	' '
English- and French-speaking countries	" "	' '
French-speaking countries and Switzerland (guillemets)	« »	‹ ›
German-speaking countries (guillemets)	» «	› ‹

The **guillemet** is separated from the word with an empty space.

The **apostrophe** is attached to the word without a space. The subsequent word spacing is shortened.
Note: avoid incorrectly using the inch sign (fig. 39) as an apostrophe (fig. 40).

The **hyphen** is used only for word divisions, for combining words or when parts of a word are omitted.

The **dash** is used with a space before and after (fig. 41). When the dash (also m-dash) means "until" it is used with small spaces.

Brackets are separated from the rest of the text by a small space.

Mathematical signs (→ pp. 64, 69, Arabic numerals) are to be separated from the numeral by a small space (fig. 42).

Table numerals are of uniform width (→ p. 75, width values) and are used when columns of figures have to be arranged one below the other as in a table setting (fig. 43). In continuous text, **normal numerals**, **lowercase numerals** or even **small caps numerals** are used.

Footnote symbols, identified in the setting by superscript numerals or asterisks, are set with a small gap between them and the text (fig. 44). The footnote symbol is also slightly distanced from the text after it in the footnote text.

The **degree symbol** is placed with a small gap after the numeral.

The **slash** is centred within small spaces (fig. 45).

The **per cent sign** and the **one tenth of one per cent sign** (→ p. 71) are separated from the numeral by a small space.

Punctuation marks including **exclamation marks**, **question marks**, **colons** and **semicolons** (→ p. 71) are either not separated from the preceding text, or a small space is used.

The **empty space** (also "**blank**") is used for the gaps between words in a continuous text. It is available in half, quarter and eighth widths (→ p. 83, em).

Page layout

Jan Tschichold says of the **type area**: "The correct position of a piece of typesetting, however successful it may be, on the paper is just as important as the careful execution of the setting itself! Clumsy positioning can ruin everything."[12] The type area describes the area occupied by continuous text on a double page.

A successfully organised type area, as well as having purely rational advantages in the design phase, also guarantees financial advantages, especially at the pre-press, plate-making and production stages (→ ch. Production, p. 167, composition).

[12] Tschichold, Jan: Treasury of Alphabets and Lettering (1995)

There are various ways of looking at the construction of type areas relating to structure, aesthetics, psychology etc. Here the type, purpose and scale of the printed matter are the factors determining the individual parameters.

The **running head** (fig. 51) and the footnotes are included when fixing the type area. **Page numbers** and **marginalia** (fig. 53), on the other hand, are not part of the traditional fixing of the type area. But all the elements involved are part of the whole.

When devising a book or other double-page jobbing work[13] (a term for small-scale commercial printed matter), the right- and left-hand pages are considered as a pair, as a design unit.

The type area is usually arranged in the same way throughout the publication (→ ch. Design, pp. 42, 43, The design grid). The unprinted areas fulfil an aesthetic function here and provide the necessary repose for the eye (→ pp. 60, 61, Perception and reading behaviour). In the classical sense, the ideal variant for the proportional ratio is the **Golden Section** (fig. 46: Area proportion in the "Golden Section", devised by Jan Tschichold).

On a double page, the outer edges or outer **margins** (fig. 47) are usually wider then the inner edges or **gutter** (fig. 48). This is because the gutter is mirrored visually when a double page is opened and thus doubled.

Tip: *To work out the dimensions, the width of the outer edges is roughly the sum of the inner edges.*

The **folio head** is the typographical term for the **page number** or **chapter number**. A folio head is the term for a page number in isolation. A running head is a page or chapter number with text added about the author, keywords etc.

The type area is described in the following **technical terms**: outer margins (fig. 47), gutter (fig. 48), footer (fig. 49), header (fig. 50), column title (fig. 51), page number (fig. 52) and marginalia (fig. 53).

Headings (headlines, titles, rubrics) (→ p. 62) provide brief information about the content, and structure the text. They should be striking, and stand out from the rest of the text.

13 The term jobbing originated because 15th century printers were mainly concerned with producing books, and so "jobbing" work was only performed occasionally. The Roman Catholic Church's letters of indulgence from this time are presumed to be the oldest items of jobbing printing and the oldest printed forms.

Headings are sometimes replaced by a single word set differently at the beginning of a paragraph. This is set with a visible gap (1 **em**[14]) between it and the following text and without other markings. Bold or italic characters are often used to make it stand out.

A text is divided into paragraphs to make it more readable and comprehensible. Here the **indent** is an important instrument for identifying new paragraphs and structuring the text.

Tip: *The indent should be clearly perceptible, but not too big. In the early days of printing the paragraph sign (also called pilcrow or alinéa) was used rather than an indent to identify a paragraph.*

A **widow** (fig. 54) is the last, not quite complete line of a paragraph if this appears as the first line in a column. An **orphan** (fig. 55) is the first line of a paragraph if this appears alone at the end of a column or a page created by a manual break.

---

[14] A term from hot metal setting, it was a hand-set space with a square cross-section of the same width as the shank.

Typeface Design

Type cannot actually be reinvented. Only the external form and the basic framework,[15] can be modified (→ pp. 60, 61, Perception and reading behaviour).

When **designing a typeface**, the rules are very clearly defined and do not give the designer much scope for eccentricity. Clarity and reticence are essential, or as Adrian Frutiger put it: "If you remember the shape of the spoon you ate your soup with, there was something wrong with it … A typeface should be such that readers do not notice it … a good typeface is both banal and beautiful at the same time."[16]

When we read a text we take in and process large quantities of information. Readers should be able to decode this information in the best possible way.

Sheer will to design is not enough for developing a text typeface. There must be a clear, lucidly ordered structure behind the **appearance of letters**, words and sentences in a typeface design. Jan Tschichold writes: "As well as the essential rhythm, it is above all the markedly clear and unmistakable form, the highly sensitive correct relationship between assimilation and dissimilation in the individual letters, in other words the similarity of all the letters and the simultaneous differentiation between the individual letters that guarantee complete legibility."[17]

If a typeface has too vigorous a life of its own it can inhibit fluent reading. For reading to be as fluent as possible, basic technical skills and also knowledge of the interplay between **visual perception phenomena** (→ ch. Design, p. 24, Illusions) are needed when developing a typeface.

[15] Not to be confused with the internal structure of a letter.
[16] Frutiger, Adrian: Eine Typografie (1981). Not available in English.
[17] Tschichold, Jan: Die Bedeutung der Tradition für die Typografie (1964). Not available in English.

Basic rules of typeface design

1. **Aims.** Decide on the purpose, design direction and technical requirements.

2. **Symmetry and differentiation.** A typeface is a homogeneous entity made up of different characters; it must present a uniform appearance in formal terms; this includes rhythmic details, equal end strokes (→ p. 68, fig. 23), visually equal stem thicknesses, similar bowls etc. But the basic forms should not be too similar, so that they are easier to distinguish: the right balance has to be found.

3. **Basic framework of the letters.** The letters "H" and "n" convey the essential qualities of a typeface. They determine the height of the face and its various proportions, such as stroke thickness, width, bowls etc. (→ p. 68, fig. 23).

4. **Weight of the letters.** The weight for lowercase letters is respectively lower than for uppercase letters.

5. **Width of the letters.** The "m" may not be a double "n"; the counter of the "o" must be the same as the optical width of the "n"; the "u" should always be a little narrower than the "n".

6. **Round forms.** Letters with horizontal curves (e.g. "o" and "n") should come above or below the imaginary type lines (fig. 56), otherwise they look too small in the overall picture.

56

7. **Ascenders and descenders.** The descenders in a typeface should never be shorter than the ascenders. The lowercase ascenders (e.g. in "f") should always be a little higher than the uppercase letters.

57

58

59

8. **Cross bar height.** The cross bar of an "H" is at the optical, but not the mathematical centre. This applies to all letters that are divided in two horizontally such as "A","B","K","S", "x" etc. (fig. 57).

9. **The diagonal and bold upstrokes.** Upstrokes as for example in "A" and "N" (fig. 58) are thinner than the downstrokes, so that they create the same optical effect. Diagonal bars that end freely (e.g. "z") taper towards the intersection point. The lines in the cross of the "X" are not taken through as straights, but offset in relation to each other; they also run slightly conically towards the intersection point. Horizontal bars are always drawn somewhat narrower than the stems (fig. 59), so that they look the same optically.

10. **The character set.** A typeface includes more characters than the 26 letters of the alphabet; numerals and other groups of figures have to be adapted to the overall picture (→ p. 91, glyphs palette).

11. **Spacing a typeface.** Individual letters do not yet make a sentence. Noticeable holes and unduly narrow spaces between letters and words make it difficult to identify words and sentences and thus make reading less fluent. Spacing means creating spaces between letters that look the same – and aligning them with the typeface size (→ pp. 57, 59), character width (→ p. 75, fig. 25), letter counters and stem weight (→ p. 68, fig. 23). Spacing is primarily done through the adjustment of the letters "H", "n", "o" and "O" (→ p. 88).

12. **Kerning a typeface.** The more carefully the adjustment has been done and the form of the characters devised, the less kerning will be needed. But kerning (→ p. 89) is always essential for creating similar-looking white areas (→ p. 75) between certain pairs of characters. A key point for kerning: only the most important character pairs (e.g. "AT","AV","Av","AC","DY","FA") should be kerned (fig. 60).

60

Spacing and kerning

As already mentioned above, noticeable white space, holes or unduly narrow distances between letters inhibit the recognition of words and sentences and make reading less fluent. Optically even distances between letters, matched to the typeface size, character width and stem thickness are thus required if a mere sequence of letters is to be made into a text with legible characters. In contrast with the larger spaces between words, this guarantees that the structure of a word can be grasped (→ pp. 60, 61, Perception and reading behaviour). This is why letters and characters are "spaced".

Tip: *For a normal sans-serif typeface the distance between characters is about the weight of a stem. Here the small "i" is ideal for measurement. If the letter counters are larger (→ p. 68, fig. 23) the set width should be somewhat greater. Typefaces in small type sizes (→ pp. 57, 59) need proportionally more space between the characters than typefaces in larger type sizes (fig. 61: determining the distance between characters).*

61

Tip: *The word space depends on the cut of the type, and on its nature and size. The minimal word space is the weight of the small "i" or the counter of the "n". The larger the letter counter, the greater the word spaces. Wider typefaces thus need bigger spaces because of the wider bowl (figs. 62, 63).*

62

63

The actual difficulty of spacing lies in the fact that the spaces between the letters cannot be mathematically identical, but simply have to look the same. If all the letters had the same spacing values, the type would seem very restless to the reader, because the shapes of letters, serifs and cross bars create powerful optical illusions.

Basic spacing

In terms of simplification according to basic forms (→ p. 67,
figs. 19–21) and visual perception, the left- and right-hand
halves of the letters are divided into groups and spaced corresponding to these. So the capital "D" is in the "H" group
in terms of the space before it, and in the "O" group in terms
of the space after it.

■	BDEFH	HINM	■
●	CGOQ	DGOQ	●
▲	ATVWY	ALTVWY	▲
■	bghijklm	adfghim	■
●	cdegoq	bceop	●
▲	vwy	rvwy	▲

Tip: *To check spacing, all the characters are placed between
uppercase "H", uppercase "O", lowercase "n", lowercase "o"
and the numeral "0".*

HAHBHCHDHEHFHGHIHJHKHMHNH
OAOBOCODOEOFOGOHOIOJOKOLO
nAnBnCnDnEnFnGnHnInJnKnLnMnNn
oAoBoCoDoEoFoGoHoIoJoKoLoMoNoOo
nanbncndnenfngnhninınjnknlnmnnnon
oaobocodoeofogohoıoıojokolomonooopo
0A0B0C0D0E0F0G0H0H0I0J0K0L0M0N0

As spacing letters is not enough to achieve a harmonious set width for the type, the process known as "kerning" is also used. **Kerning** (undercutting) means changing the spacing laid down in traditional spacing for certain pairs of letters.

The **kerning value** is usually negative. For the "**kerning pair**" "AW", for example, the width of the "A" is reduced (fig. 64). When kerning, negative kerning values are fixed for the most important characters; this allows the width of the second character to be absorbed into the width of the first character.

The **kerning values** are fixed in Long or Short **Kerning Tables** or **Aesthetics Tables**, and added to the space in the appropriate letter combinations.

Italics

The average slant on italic scripts is 12 degrees. Angles of inclination of less than 10 degrees (fig. 65) scarcely look any different from the upright variant. The face version collapses visually at more than 16 degrees (fig. 66).

Tip: *Anyone wanting to develop professional typefaces can use systems such as Fontlab or Fontographer (→ p. 96, Tips and links). Please also refer to* ch. Law, p. 271, Typeface Protection Act.

Font Technology

Font formats

Technologies, programs (→ ch. Digital Media, p. 113, Software) and font formats determine the extent to which typefaces can be used. Today's fonts are digital (→ pp. 63–67, Origins of type). Characters are defined mathematically according to their outlines and supplied with **hints**, which improve the appearance of type on low-resolution devices (→ p. 95, hinting).

Hints were first introduced with the **PostScript®** page definition language developed by Adobe Systems. All that was available before this development were **Bitmap fonts**. Each letter in a Bitmap font (fig. 67) was represented as an accumulation of pixels (→ p. 93, pixels) in a matrix on the screen or in a printout. A separate font was required for each typeface size.

67

The **font** (also font file) defines the digital implementation of a certain character set, in a particular variant and file format. Hints ensure that the letters make as good a visual impression as possible at low resolutions.

The **Adobe PostScript Type 1 format** (standard for digital font types; International Organization for Standardization, Glyph Shape Representation, ISO 9541) describes all the characters in a character set as an abstract mathematical outline using **Bézier curves** (fig. 68).

Anchor points are the points fixing **Bézier curves**. **Vectors** determining the direction and degree of the curves originate from these points. A Bézier curve needs at least two anchor points, the start and end points.

68

The introduction of the Type1 format made it possible to represent and issue type in any size at all.

Apple developed the **TrueType format** in the late 1980s as an alternative to Adobe's Type 1 standard, and later licensed it to the Microsoft Corporation. While the "Adobe outlines" consist of cubic Bézier curves, TrueType defines an outline using quadratic equations ("Quadratic B-Splines"). Unlike PostScript, which uses a Bitmap character set for on-screen display and PostScript Setting for printing,

69

TrueType uses a description only for screen display and printing. TrueType features mainly outside the "Mac world".

The **OpenType font format** is the first font format that can be used **cross platform**. OpenType was developed jointly by Microsoft and Adobe. This format is supported by all current operating systems without any additional software, which greatly simplifies font management. The following figure shows the glyphs palette in OpenType.

OpenType is based on **Unicode character coding** (international 2-byte character coding). This means that the usual assignment restrictions – PostScript fonts were restricted to 256 glyphs – no longer apply. OpenType fonts can contain over 65,000 glyphs. OpenType fonts also contain all the signs available in a character set (e.g. small caps, lowercase numerals and ligatures) in a single font file.

Glyphs — Show: Entire Font

Ł	ł	Š	š	Ý	ý	Ž	ž	¦	–	×	!	"	#	$	%	&	'	(
)	*	+	,	-	.	/	0	1	2	3	4	5	6	7	8	9	:	;	<
=	>	?	@	A	B	C	D	E	F	G	H	I	J	K	L	M	N	O	P
Q	R	S	T	U	V	W	X	Y	Z	[	\	]	^	_	`	a	b	c	d
e	f	g	h	i	j	k	l	m	n	o	p	q	r	s	t	u	v	w	x
y	z	{	|	}	Ä	Å	Ç	É	Ñ	Ö	Ü	á	à	â	ä	ã	å	ç	é
è	ê	ë	í	ì	î	ï	ñ	ó	ò	ô	ö	õ	ú	ù	û	ü	°	¢	£
§	ß	®	©	´	¨	≠	Æ	Ø	±	≤	≥	¥		æ	ø	¿	¡	«
»	…	À	Ã	Õ	Œ	œ	–	—	"	"	'	'	÷	ÿ	Ÿ	€	‹	›	fi
fl	·	‚	„	‰	Â	Ê	Á	Ë	È	Í	Î	Ï	Ì	Ó	Ô		Ò	Ú	Û
Ù	ı	ˆ	˜	¯	˘	˙	˚	¸	˝	˛	ˇ	Ć	ć	ĉ	Ĉ	Ċ	ċ	Č	č
Ď	ś	Ś	ř	Ř	Ŕ	ŕ	Ŝ	ŝ	ź	Ż	ą	Ą	ă	Ă	ā	Ā	Ē	ē	ĕ
Ĕ	Ė	ė	Ę	ę	Ě	ě	Ĝ	ĝ	Ğ	ğ	Ġ	ġ	Ģ	ġ	ň	Ň	Ń	ń	Ź
Ş	ż	ŷ	Ŷ	ŵ	Ŵ	ū	Ũ	ŭ	Ŭ	ū	Ū	Ů	ů	Ű	ű	Ų	ų	Ĥ	ĥ
Į	į	İ	ĭ	Ĭ	ĩ	ī	Ī	Ĩ	ő	Ő	ŏ	ō	Ō	Ĵ	ĵ	ħ	Ħ	ď	Đ

Characters from other languages (e.g. Eastern European)
or other extended character sets can also be defined in this
single font file. OpenType fonts support the so-called intelligent typographic functions of modern layout programs
such as automatic character substitution for ligatures or
decorative letters.

Tip: *The convenient OpenType functions now work for Adobe
InDesign, Adobe Illustrator and Adobe Photoshop as well. In
applications that do not actively support OpenType (e.g. in
older versions), OpenType font formats can be used just like
other font formats even though the OpenType layout features
are not available.*

Character coding defines a process in which the general
representation of a sign (e.g. of a letter, a numeral or a symbol) is displayed simply by means of a code (e.g. ASCII or
Unicode). Character sets are based on so-called character
codings.
   ASCII (American Standard Code for Information Interchange) is a 7 bit binary code for the standard representation
of up to 128 letters and signs, later also using 8 bit numbers,
making it possible to represent 256 characters. Unicode presentation with 16 bit numbers has started to gain acceptance
recently. It makes it possible to display over 65,000 different
characters.
   Multiple Master (MM) fonts are based on the PostScript
Type 1 format. Unlike the limited number of individual cuts
within a type family, the Multiple Master technology allows an
unlimited range of type variants to be created in a typographically correct (undistorted) manner and with infinitely variable
adjustment.

Type management

With Mac OS X, PostScript Type 1, TrueType and OpenType, fonts can by used without additional tools. Mac OS X can even handle coded TrueType fonts coded for Windows. These are not available in the so-called Classic environment, however, so a type management program is still needed when using PostScript Type 1 fonts here. The fonts are installed by placing font files in special font folders.

Hint: For additional information see p. 96, Tips and Links.

Fonts on a display screen

The technical features of a monitor (→ ch. Digital Media, p. 113) greatly restrict the quality of font presentation. In particular, they make it difficult for information to be absorbed without a lot of effort (→ p. 61, legibility). The grey value (→ p. 77) that is so important for legibility can also be severely compromised.

   The problem arises because of the monitor's low resolution, on average 72 dots per inch (dpi). In addition, there is the monitor glare effect (→ ch. Digital Media, p. 113) and the fact that type on a monitor is unstable and can also be affected by movement. The display quality of fonts on the monitor also depends on other factors such as font format, the quality of the font drawing, of the software and the particular operating system.

   Fundamentally, letters are displayed on screens as **pixels**. These do not allow defined curves – the **outlines** – to be displayed precisely. This does not particularly affect large forms, but is important for small font sizes: there are not enough pixels available to display these shapes correctly. The smaller the font size, the less of the font's character shows up on screen. For this reason, for screen design the basic type should not be smaller than 12 pt when displayed at 100 per cent (→ pp. 57, 59, type size).

70

Anti-aliasing can considerably improve the way a font displays on screen. Anti-aliasing is a process used to smooth unattractive, step-like edges when displaying letters or objects on a low-resolution monitor. It works by adding colour shades to the edge of the letter (outline) and the background colour. The outlines of the letters appear much smoother, but also less sharp (fig. 70).

ClearType is a process developed by Microsoft used to create the clearest possible type image on computer screens. The main tool used here is anti-aliasing. ClearType is supposed to improve the legibility of small typefaces on coloured liquid crystal screens in particular; these screens are used for laptops and are intended to be used for **e-books** (electronic books).

71

The following formal qualities of a typeface are the key to showing material on **screen** appropriately or in a reader-friendly way: high x-heights, large internal areas and open forms as well as even type thicknesses, as in the Verdana typeface (fig. 71).

Tip: As a rule, sans-serif faces are most suitable for this. But if serif faces are used, they should be structured simply and robustly (e.g. slab-serif) (→ p. 74, fig. 5). Fonts with fine lines, large differences in type thickness and diagonal variants are not recommended for use on screen.

*Tip: The **set width** (→ p. 75) on the screen should not be too narrow, as word images can blur. The distance between lines (→ p. 75, Line spacing) should also not be too small. Typographical rules should also be followed for monitor typography, e.g. the correct opening and closing inverted commas should be used, not inch signs (→ p. 80). Justified setting should also be avoided (→ p. 77) as well as word divisions.*

The monitor technology's inability to display existing typefaces adequately has meant that system and also font producers have started to design special typefaces for monitors – so-called screen fonts or web fonts – or to optimise existing fonts for the screen (hinting).

Hinting allows font outlines to be optimally displayed on low resolution devices without changing the typeface design (fig. 73).[18] In hinting, mathematical instructions (→ p. 90, hints) are embedded in the fonts so that the outline of a letter better matches the screen matrix of the display medium (fig. 72).

Tip: Not all hints are the same. Many tools contain automatic hint algorithms. These are generally better than nothing, but they cannot replace custom-designed hints. Creating manual hints is a laborious process, however.

Hint: See also ch. Law, p. 271, Typeface Protection Act.

With hinting   Without hinting

---

[18] Ever-increasing printer definition in recent years has meant that grid image problems have become minor for paper documents, but they are still significant in relation to monitors.

Tips and Links

Fundamentals

Friedl, Friedrich: *Typography. An Encyclopedic Survey of Type Design and Techniques throughout History.* Black Dog & Leventhal Publishers, New York (1998)

Font technology

Karow, Peter: *Font Technology. Methods and Tools.* Springer Publishing-Telos, New York (1994)

Adobe: www.adobe.com/type/topics
Decodeunicode: www.decodeunicode.org
Fontlab: www.fontlab.com
Microsoft: www.microsoft.com/typography
Unicode: www.unicode.org

Font management

www.apple.com/macosx/features/fontbook
www.extensis.com/products/
font_management
www.linotype.com/fontexplorerX
www.microsoft.com/typography

Lettering

Cheng, Karen: *Designing Type.* Yale University Press, New Haven (2006)

Frutiger, Adrian: *Signs and Symbols. Their Design and Meaning.* Watson-Guptill Publications, New York (1998)

Kapr, Albert: *The Art of Lettering.* K. G. Saur, Munich (1983)

Tschichold, Jan: *Treasury of Alphabets and Lettering. A Source Book of the Best Letter Forms of Past and Present for Sign Painters, Graphic Artists, Commercial Artists.* W. W. Norton & Company, New York (1995)

Font labels:

Die Gestalten: www.die-gestalten.de/fonts
Emigre: www.emigre.com
House Industries: www.houseind.com
Lineto: www.lineto.com
Linotype: www.linotype.com
Optimo: www.optimo.ch
Typotheque: www.typotheque.com
Underware: www.underware.nl

Magazines and forums

Baseline: www.baselinemagazine.com
Communication Arts: www.commarts.com
Typophile: www.typophile.com

Organisations and events

Association Typographique International (ATypI): www.atypi.org
St Bride Library: www.stbride.org
Type Directors Club: www.tdc.org
TypeCon: www.typecon.com
Typo Berlin: www.typo-berlin.de
TypoTechnica: www.linotype.com

Other

Identifont online font directory: www.identifont.com

5.0

Life and work without a computer? To most designers, this sounds inconceivable. Whether for a first draft or for post-production, at some point or other, we invariably rely on this virtual extension of hand and mind. In digital media – a field which encompasses everything from websites and DVDs to interactive games and audiovisual installations – the computer takes on an even more prominent role. Rather than only being used as a flexible design aid, the computer becomes a communication tool, medium and platform in one.

In this chapter, we sneak a peek below the surface of today's bold, bright and often moving digital imagery. Like the print world, the digital world is all about communicating, highlighting and visualising information. In addition to a strong creative knack, working with digital media also calls for an impressive set of interdisciplinary skills. These skills include basic knowledge of programming rules and information architecture as well as informed awareness of the opportunities, compatibility issues and restrictions that come with the multitude of formats, programs and carrier media in use today.

Over the following pages we will clarify the genre's technical jargon and give you a few hints on how to work with soft and hardware, file formats, platforms, networks and animations. We'll guide you in your use of information, programming and digital standards to achieve the aesthetics you desire and help you to wow the crowds and make your mark with coherent, elegant solutions. In addition, this chapter devotes plenty of space to frequently neglected usability issues such as clear processes, loading times and interactivity – aspects often forgotten in the rush of creative abandon.

Print aficionados and those who prefer a more "traditional" approach to design would also do well to check out this insightful overview of current digital design nomenclature and methodology. The latest techniques will help you optimise your work for the internet or use new trends such as tagging and social software to your advantage. Welcome to the digital playground; it's yours to design!

Digital Media

5.1	**File Formats**	102
5.2	**Hardware and Software**	111
	Hardware	112
	Software	113
5.3	**Internet, Networks and Programming**	114
	Internet trends and developments	116
5.4	**Moving Pictures**	119
	TV standards	119
	High Definition format	120
	DVD	123
	Compression	124
	Web and computers	124
	Software	125
5.5	**Disciplines and Workflow**	126
	Grid systems for digital and dynamic media	129
5.6	**Web Design**	133
	Programming / The source code	135
	Basic framework of an HTML file	138
	Other programming languages	141
	Barrier-free websites	141
	Usability	142
5.7	**Presentation and Structures**	144
5.8	**Tips and Links**	146

File Formats

A **file format** or file type determines how computer data are saved. The format defines how the data are coded and interpreted when called up or stored. Renaming a file changes the filename, but not the file format. Even though it is not quite correct to do so, the terms **data format** and file format are frequently treated as synonyms. Every file format is at the same time a data format, but not every data format a file format. Thus, for example, a data format can consist of several files in different file formats.

    The format of a file is indicated by the file **suffix**, the **section** at the end of the filename. A filename is usually structured and identified like this: **filename.suffix**. Depending on the operating system, a suffix often contains no more than three characters.

Tip: Suffixes are not relevant when saving material on Macintosh computers, as Macs recognise files by other properties. But the suffixes are still important, above all for exchanging data over a number of platforms.

Not every program recognises all file formats. Thus only some of the various formats can be imported or exported in certain programs. Formats that can be read by only one program cannot be used for exchanging data between more than one program.

    An **exchange format** is usually a data format that can be used in several programs and/or on one or more computer platform. The first formats of this kind were essentially simple in their conception. The .txt format, using only so-called ASCII[1] characters, is still an important format today, however.

Layout file formats are not generally available as exchange formats (exceptions: → pp. 104–109, table), unless the recipient has the program in question.

---

[1] Abb. for American Standard Code for Information Interchange.

Tip: The graphics and images embedded in layout files are not usually saved in the layout file itself. The layout file contains only references to the embedded graphics and images. Hence graphics and images embedded in the original files should not be deleted or relocated, as otherwise this will trigger a files missing message. Some programs offer the possibility of embedding image data in the layout. Note that when using this option the layout data file will be considerably larger in size.

Note: As a rule, "new" program versions cannot be opened in "older" versions!

Tags (→ pp. 117, 136, 137) are **data bytes** in which a particular length in the data blocks contains the image information. Tags also contain information on image dimensions or resolutions (→ pp. 108, 109, TIFF). Container formats are files that contain a whole range of different data, but are allocated to a certain file category by a uniform extension.

The following summary shows current file formats and their file suffixes, and lists their particular features.

Format type	File suffix	Name
Layout file format, vector-based graphics data format	.ai	Adobe Illustrator (AI)
Audio file format	.aif	Audio Interchange File Format (AIFF)
Multimedia file format (audio and video data format)	.avi	Audio Video Interleave (AVI)
Pixel-oriented picture file format	.bmp	Windows Bitmap
Graphics file format	.dcs	Desktop Color Separation
Text file format	.doc	Microsoft Word (DOC)
Vector-based data format	.dwg	AutoCADdrawing
Graphics data format, picture data format	.eps	Encapsulated PostScript (EPS)
Program data format	.exe	Portable executable
Graphics file format	.fh6	Macromedia Freehand Version 6.x
Multimedia data format	.fla	Flash
Multimedia data format	.flv	Flash video
Pixel-oriented picture file format	.gif	Graphics Interchange Format (GIF)
Compression file format	.hqx	File compressed with BinHex (Mac)
Text file format	.htm .html	Hypertext Markup Language

Exchange and use	Notes
Suitable for print and non-print production.	Adobe Illustrator's own program format. Variant on the EPS format.
Can be used across platforms. Suitable for non-print production. Not suitable for print production.	The standard coding for audio files is a sampling frequency of 44.1 kHz and a data depth of 16 bits and stereo sound (CD-Standard). Not compressed. Large volume of data.
Can be used across platforms.	Audio and video data are stored in the same file.
Suitable for neither print nor non-print production.	Standard Windows format. The format can be compressed without data loss. Large volume of data.
Suitable for print production.	The format is a special form of the EPS format. Storage of CMYK or special colour data in separate files; or in one file as of version 2.0.
Can be used across platforms. Whether Word formats are understood on import into a program depends essentially on the import function of the program concerned.	As a rule the format saves formatting such as characters, paragraphs or tables along with any text.
Is supported by most of the well-known CAD programs.	Format developed for AutoCAD software.
Cross-media readable (all programs and platforms). Not suitable for non-print production.	The format can contain vector graphics and pixel images. EPS images can be scaled without loss (except for embedded pixel graphics).
WIN 32 systems.	Executable program file (PC).
Cannot be used across programs.	Macromedia Freehand's own program format. Layout
Suitable for non-print production.	Source data from Adobe (Macromedia) Flash.
Suitable for non-print production.	Video format from Adobe (Macromedia) Flash.
Not suitable for print production.	The format can be compressed without loss using the LZW process. Small colour range (256 colours, 8 bit colour depth). Interlacing: the image is not built up line by line (from top to bottom) on the monitor but in focus stages; thus it can be recognised after half the transmission period. Animated GIF: in GIF animation programs several individual GIFs are ordered one behind the other. This format is used particularly for advertising banners and for gimmicks on websites. Can be made transparent.
	Hexadecimal compression as safeguard during binary transmission (modem). Addition to .sit compression.
Can be used across platforms through interpreter (browser). Foundation of the World Wide Web.	ASCII file. This file format can only be read and interpreted with the help of a browser.

Multimedia file format	.iff	Interchangeable file format
Layout data format	.indd	InDesign file
Pixel-oriented picture file format	.jpg .jpeg .jfif	Joint Photographic Experts Group File Exchange Format
Compression data format	.lzw	LZW compressed image file
	.mid	File format for Musical Instrument Digital Inte (MIDI)
Multimedia data format	.mov	QuickTime Movie
	.mpv	Microsoft Project File
Audio data format	.mp3	MPEG Layer3 (MPEG Standard 3) ISO MPEG I Audio Layer 3 (Moving Picture Experts Group)
Multimedia data format	.mp4	MPEG 4 (Moving Picture Experts Group)
Font data format	.otf	OpenType Font (PostScript)
Image file format	.pcd	Kodak PhotoCD
Image file format	.pict (Mac) .pct (PC)	Macintosh Picture, PICT image
Document format	.pdf	Portable Document Format
Pixel-oriented picture file format	.png	Portable Network Graphics
	.ppt	PowerPoint Presentation (Microsoft)
Document format	.ps	PostScript

	Amiga exchange format for BMP, audio.
Can open XPress files (only up to version 4).	Adobe InDesign's own program format.
As a rule a pure web format.	Full colour range. Image quality dependent on compression. All current web browsers read the format.
	Compression process by Lempel-Ziv-Welch. LZW compression is loss-free.
	Protocol for the transfer of musical control information between instruments and/or computers.
For saving video, animation, graphics, 3D and sound data.	Container format comprised of several tracks according to volume (image, sound, captions …).
Microsoft Project file	Microsoft Project's own program format.
Suitable for non-print (especially web) production. Can be used across programs with widely used, open decoders.	Compressed sound file with sophisticated compression technology. Different degrees of compression can be selected in the coding. The greater the compression, the poorer the quality.
Compression method for mobile multimedia.	
	Newest font format (also TrueType, .ttf) which – through Unicode – can contain up to 65,536 glyphs (different languages and their accents, ligature etc.).
Archiving of photo material.	Quasi standard for the professional archiving of photographs. Also standardised for digitalisation.
Not suitable for print production.	Standard Mac image format until OS9, now PDF. Can contain both vector and pixel data.
Cross-media readable (all programs and platforms).	Page description language that exactly reproduces fonts, formats, graphics etc. independently of the program or operating system being used. Quasi standard for digital exchange of data (especially prepress).
Not suitable for print production.	Open substitute for .gif format without the animation function.
Cannot be used across programs.	Microsoft PowerPoint's own program format.
Can usually be read and edited with a text editor. Requires a PostScript capable output device (hard and software). Can be used across platforms.	Page description language that contains PostScript data for print output.

Pixel-oriented image file format	.psd	Adobe Photoshop
Layout file format	.qxd	Quark Xpress
Compression data format	.rar	RAR, Roshal Archiver
Text data format	.rtf	Rich Text Format (RTF)
Compression file format	.sit	StuffIt
Audio file format	.swa	Shockwave Audio (SWA)
Multimedia data format	.swf	Shockwave Flash, Small Web Format
Pixel-oriented image file format	.tif .tiff	Tagged Image File Format
Font file format	.ttf	True Type Font
Text file format	.txt	Text only (TXT)
Audio file format	.wav	Waveform file
Vector-based data format	.wmf	Windows Metafile
	.xls .xlsx	Excel spreadsheet
Compression file format	.zip	ZIP compressed file

Not an exchange format as a rule. .psd files cannot be imported into other programs without difficulty. In order to import .psd files into a layout program, the contents of all layers must first be reduced to a background layer in Photoshop and then saved in a different format (e.g. TIFF). InDesign can import .psd image data, though all the levels and possibly transparencies are lost. Not suitable for print production.	Adobe Photoshop's own program format for interpreting graphics data without loss. Levels, objects and document settings are kept. Text can also still be edited in the text layer. Large volume of data.
InDesign can open Quark XPress files until version 4.	Quark's own program format. Not an exchange format.
Only decompression is open; compression is licensed (cannot be used in all programs).	Compression file format of the RAR program.
Any word processing program can import RTF files. Cross-media readable (all programs and platforms). Suitable for print and non-print production.	Formatted text file. The format saves formatting as well as text.
	Compression file from StuffIt (mostly Mac).
For internet use. Suitable for non-print production. Not suitable for print production.	Can be transmitted using the streaming process. Using streaming, it is possible to listen to a file while it is still being downloaded.
Suitable for non-print production (web).	Compiled Flash file, suitable for web (with Flash player).
Exchange format for pixel image data. Not suitable for non-print production.	Large data volume. Can be compressed without loss.
	Font is described through outlines, similar to vector graphics The output device is the first to substitute the contours with pixels.
Cross-media readable (all programs and platforms). Suitable for print and non-print production.	Contains no formats, colours, pixels etc. Uses ASCII.
Can be used across platforms Suitable for non-print production. Not suitable for print production.	Non-compressed as a rule. Large data volume. ADPCM: WAV variant with compression (25 per cent).
Cannot be used across platforms.	Used by Windows for temporary storage.
Cannot be used across programs.	Microsoft Excel's own program format. .xlsx format is based on XML.
Can be used across programs and platforms.	

Tip: A sound middle course for compressing a JPEG file is to use about 75 per cent. Compression takes place during the saving process. This means that where possible a JPEG file should not be worked on directly and then saved again as a JPEG. The quality would deteriorate each time it was saved, as the file would become increasingly compressed. The recommended procedure is to use the uncompressed source file, such as BMP or PSD, to work on and then make all the changes to this file.

Pixel file formats contain image data in a matrix of image dots (pixels). Each of these pixels is allocated certain coordinates and a colour value. **Bitmap** is the simplest graphics format. Here the pixels are defined within a two-dimensional system of coordinates with an X-Y value (position of the pixel) and a colour value. However, these files can become very large, especially when the picture contains a very large number of colours. A suitable compression process can sometimes reduce this data volume.

    Vector formats contain a mathematical description (→ ch. Typography, p. 90) of individual objects in an image (e.g. lines, circles, polygons). In the simplest case, a line is defined. Vector formats are easy to scale. Figure 1 shows a comparison of the way in which vector- and pixel-based image or graphics content is defined. **Metafile formats** can contain both pixel and vector data. Encapsulated means that the files are encoded in PostScript language. They can often no longer be modified after saving in a program other than the one that produced them.

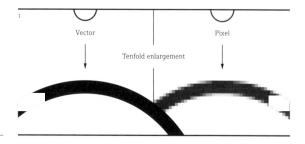

Vector   Pixel

Tenfold enlargement

Hardware and Software

The **bit (binary digit)** is the smallest **information unit** that
a computer can represent. In information technology the
value of a bit is also called a status. A bit can have only two
statuses: either 1 or 0, true or false, yes or no, on or off. It
forms the basis of the **binary numeral** system used by all
computers. The **byte** is the basic working unit, a computer's
addressable memory unit. A byte consists of **eight bits**,
which means that $2^8 = 256$ different combinations or characters are possible.

1 bit	= 0 or 1
1 byte	= 8 bits
1 kilobyte (KB)	= 1024 bytes
1 megabyte (MB)	= 1024 kilobytes
1 gigabyte (GB)	= 1024 megabytes
1 terrabyte (TB)	= 1024 gigabytes

Bit depth measures the **number** of **bits** with which a graphic
input or output device reproduces each individual pixel in
an image. The bit depth determines colour or tonal range.
An eight bit colour depth means 2^8 and makes it possible to
represent 256 colours. Computer **graphics cards**, for example, work with bit depths of up to 24 bits.

Colour depth	Colours to be displayed
16 bits	$2^{16} = 65,536$
24 bits	$2^{24} = 16,777,216$

Hardware

The term hardware refers to all the physical elements in electronic data processing. This also includes assemblies and components.

The German civil engineer Konrad Zuse (1910–1995) developed the first programmed calculator (Z1) working with a system of binary numbers, but still on a mechanical basis. He introduced this computer to the public in 1936.

The **computer** as hardware consists mainly of the **central unit** (control unit) – the heart of a computer – and the **peripherals**. The central unit includes above all the processor (CPU), the working memory (RAM) and the various bus and connection systems (interface elements, slots for extension cards). The peripherals include all the components that can be connected to the central unit such as input and output devices.

Input and output devices

In computer technology, all the components that can feed information into a computer are known as **input devices** (fig. 2). These include the mouse, keyboard, **CD-ROM drive**, **DVD drive**, graphics tablet, data glove, camera, microphone or scanner.

Output devices are all the components that make the output from all the data saved or modified in the computer visible, audible or accessible to the user in some other way. Output devices include monitors, printers, projectors, plotters or DVD burners. Data **storage media** include **USB sticks**, **external hard drives** and also CDs and DVDs.

The monitor

The **monitor** or **screen** is an output device used to display and reproduce material such as characters or images. The graphics card forms the interface between the computer and the monitor; it makes the necessary computations for monitor display. Today a distinction is made between two main monitor types: the **LCD screen**, also called **flat screen**, and the **CRT screen**, also called cathode ray tube screen.

Monitors produce all of their colours by mixing red, green and blue (fig. 3) (→ ch. Design, pp. 9, 10, additive colour mixing).

The resolution available depends on the physical characteristics of the display or pick-up device concerned. The **screen size** is given in inches, and always refers to the diagonal measurement of the screen (fig. 4). **Screen resolution** describes the number of pixels: number of lines × dots per line. Today's high resolution monitors have a resolution of 1024×768 pixels (px) and above.

Software

Programs and data are known as software. These include the operating system, system extensions, drivers, programs or program extensions. A computer's basic software is its **operating system**. This controls and organises all fundamental functions such as launch programs, managing and using storage systems, organising peripheral devices or executing programs such as **image-** or **word-processing programs**.

Internet, Networks and Programming

Computer networks are conceivable in a variety of forms, structures and sizes. Even combining a laptop with another computer creates a small computer network. A computer network is several computers linked for the purpose of data exchange or information exchange. Communication takes place via various **protocols**. The following network types can be identified, depending on size:

The **LAN – Local Area Network** – is a **local network** (fig. 5). The term describes networks that take up little space and also networks with few computers attached but that can be positioned over a wide area. One example of a LAN is an independent company or home network. Cable-linked LANs work via **Ethernet** in most cases. A wireless local network is known as a **WLAN (Wireless Local Area Network)** or Wireless LAN.

5

But the wide availability of WLANS has also led to set-ups used by a number of people that formally speaking are not pure LANs, but a common internet access system.

A very large network, like for example several local networks that can be connected over large distances, is known as a **WAN – Wide Area Network**. Data within a network is conveyed via network cables, network cards, network protocols and various network components such as routers and switches.

The **internet** is the name for the world-wide electronic network. Almost all networks, including the internet, work with the **TCP/IP** protocol (Transmission Control Protocol/ Internet Protocol), which controls addressing and data exchange between various computers and networks in the form of open standards. The **Domain Name System (DNS)** is an important part of the internet infrastructure. In order to be able to address a particular computer, the IP protocol identifies the DNS with an unambiguous IP address. In this way, an internet address is translated into a corresponding IP address and the data packets are addressed to or by the desired recipient.

E-mails are electronic messages or communications transmitted and exchanged via the internet. **FTP (File**

Transport Protocol) is used to transfer files between computers, via the internet, for example.

Instant message systems provide live chat capabilities and often also make it possible to share images via web cams and data over networks.

A closed computer network based on internet technology is called an **extranet**. It is comparable with a **Virtual Private Network (VPN)**.[2] Extranets, unlike intranets, can also be accessed from the outside; but unlike the so-called public network they allow access only to registered users. Most LANs are configured as an extranet of this kind.

The term "internet" is often used as a synonym for the **World Wide Web** (WWW or web for short). But the WWW is only one of many **applications** of the internet. The WWW is based on a hypertext-supported information and source system for the internet: **Hypertext Transfer Protocol (HTTP)**. HTTP is used mainly to load websites and other data from the World Wide Web into a web browser.

The WWW is based on three core standards: HTTP as the protocol by which the browser can request information from the web server; HTML as the document description language that determines how the information is structured and how documents are linked (**hyperlinks**); URLs as unique addresses or designations for a resource (e.g. a website) used in hyperlinks.

A **web browser** makes it possible to navigate on the web. **Bookmarks** make it possible to re-access pages or files that have been visited on the internet. All current web browsers permit storage of WWW addresses. **Cookies** are information files sent by the web server of the site being accessed. They allow a web server to deposit information in the form of text files on the computer that is making contact so that it is available when the next contact is made. As a web server has no direct access to the user's storage medium, the browser has to be used for this purpose.

Advantage: If a site is called up again, the data relevant to it are transferred to the web server from the cookie file. That means that data such as a user's e-mail address and name are transmitted to the web server and do not have to be inputted again.

Expansion of an intranet; joins two or more intranets via a general internet connection.

Disadvantage: Cookies can also be used to build up user profiles (surfing habits) and for ferreting out information ("the transparent customer"). Some web servers permit use of the material they offer only if cookies are allowed.

Tip: *It is recommended that cookies are usually disabled in the browser and activated temporarily only for trusted sites. If cookies are enabled, great care should be taken when inputting data* (→ p. 133, web design).

Internet trends and developments

Web 2.0 is a new step in the distribution and use of the internet. The term **Web 2.0** – also second generation web – was coined in the year 2004. The Americans **Dale Doherty** and **Craig Cline** are seen as the inventors of this catch phrase, though actually they were only looking for a name for a conference on new trends and techniques on the internet. The American publisher Tim O'Reilly has addressed Web 2.0 with particularly verve, and has published numerous articles and essays about it (→ p. 146, Tips and Links).

Web 2.0, unlike the usual software numbering, does not mean that there was a new version of the World Wide Web at a particular time. Nothing fundamental happened to the basic principles by which the internet itself functions. In fact, Web 2.0 describes a new approach to use. This was made possible by new techniques and applications, in the private sphere as well. The key feature of this development is that the traditional boundaries between providers and consumers of material are disappearing. The basis for this is a series of innovative technologies, some of which are explained below.

Feeds, so-called **subscription services**, mean that users can be informed automatically of new material on a website. For example, in the case of news services users do not have to check the website for new material themselves. Feeds reverse the usual information paradigm: instead of downloading the information, it is supplied to the user automatically. The main instruments here are so-called **RSS feeds** – Really Simple Syndication. These are XML-based files that create themselves automatically very easily, and can be integrated into other applications.

Tagging is the name for a technique that users can deploy to mark websites or individual contributions they have read with terms they can choose at will, in other words they can place **tags** (→ pp. 103, 136, 137). Tagging extends the classical **keyword** strategies (→ ch. Marketing, p. 228) and makes it possible to categorise data very precisely and comprehensibly.

Permalinks (a contraction of permanent links) are used mainly in weblog systems. They make it possible to create a permanent link with individual articles or versions of articles on a website, e.g. **URI** (Uniform Resource Identifier), or of an individual weblog entry. The advantage is that an article can be found again, even if the address or the content has been changed.

Permalinks make permanent linking of internet resources possible, conversely so-called **trackbacks** provide authors with information about who has linked up with their websites.

The technologies function independently of location and with any internet-compatible terminal device, in other words independently of devices and medium. There are hardly any hard or software-related barriers to their use. Undue graphic profusion has largely been suppressed by general, recognised use schemes (→ p. 142, barrier-free access).

A further aspect of the Web 2.0 philosophy is the Open Source Model. Systems are opened by the use of open technologies, so that they can be developed further or combined by different developers. One consequence of this is that all components are constantly in development.

APIs and **mashups**. Open programming interfaces have long been used in software development to link different systems. Large service providers such as Amazon or eBay have revealed the interfaces to their databases and exploit the creativity of a world wide user community for use by millions – and thus gain advertising for their services. The term **mashup** was introduced for the resultant combinations of existing web content and web services. Examples of this are the linking of geo-data on Google maps with other content such as meta-information.

AJAX (**Asynchronous Javascript and XML**) is a programming technique that first broke down the traditional interchange between user activity and server processing on the web. AJAX makes it possible for subject matter and objects to be reconstructed on the browser without having to reload the entire page content after every action. The browser responds considerably more quickly, thus creating the impression of working with the user interface on a traditional desktop application.

Older technologies often need longer loading times. Reconstructing the complete page creates a noticeable wait for website users.

The components described are directed in the first place at allowing every user to benefit from the "wisdom of the many". Good examples are Wikipedia and MySpace. The terms **social software** and the **social web** are often used in this context. The "intelligent web" develops almost of its own accord in this way – simply through adoption by the broader user community and without having some architecture cobbled together in the lab imposed on it.

Moving Pictures

Moving pictures is the name for a rapid sequence of individual images that viewers perceive as continuous movement. About **24 frames** per second are sufficient for the human eye to create the **illusion** of continuous movement harmoniously, so long as the individual images do not differ from each other too greatly. **Frames per second (fps)** identifies the image change frequency: in other words the number of images recorded and displayed per second. Frame means image or picture in this context.

A fundamental distinction is made among moving picture material for **TV/video**, **DVD** and **web/computer**. For television the pixels are rectangular, but for the computer world they are **square** (→ p. 120). This leads to different figures for picture dimensions: the 720×576 rectangular pixels in a **PAL TV** image correspond to 768×576 square pixels on a computer monitor. Frequently software will rescale automatically (e.g. Quicktime Player, Photoshop CS).

TV standards

There have been scarcely any changes since colour television was introduced in Germany in 1974: in the **PAL** regions, like for example Europe, Australia and Africa, the image is transmitted at 50 Hz with 720 pixels in 576 lines. In regions that broadcast in NTSC, like North America and Asia, the standard is 720 pixels in 480 lines at 60 Hz.

This format, known as **SD (Standard Definition)**, is broadcast to television sets in the so-called **interlaced mode** – relating to each country's power supply (Hz) in 50 (PAL) or 60 (NTSC) fields. There are essentially two regional broadcasting formats within this TV standard, PAL and NTSC. The French variant **SECAM** need not be taken into account because it is so little used internationally.

Format	Resolution	Frame rate	Distribution
PAL	720×576	25 fps	Europe, parts of Asia, Africa etc.
NTSC	720×480	30 fps	USA, Japan etc.

Current **SD recording formats** are: **Digi-Beta**, **Beta SP** (analogue) and **DVCPro50**. **Mini-DV**, **DV Cam** and **DVCPro25** also feature in the so-called amateur and semi-professional sphere.

High Definition format

High Definition (HD) is becoming increasingly accepted in **Europe**. HD has been established for a long time in the **USA** and in particular in **Japan**, where legislation has also considerably speeded up the introduction of HDTV. At the time of writing, two **HDTV** formats are available: **720p** and **1080i**. For 720p the resolution is 1280×720 pixels, so the **full frame** is always transmitted. All picture lines are thus shown at the same time; the "p" after the figure (720p) stands for **progressive**. The HD format increases the resolution by four to five times over standard definition (→ p. 119, SD). Unlike SD, HD works exclusively with **square pixels** – just like the computer. Here too there are two different standards with two different frame rates in each case.

The **1080i** format increases the resolution to 1920×1080 pixels. However, 1080i works in an **interlaced mode** like **SDTV** hitherto. The word interlaced is used because a picture is not shown in full, but two fields are used, one following the other, and the format is called 1080i. The following table compares the TV formats.

TV formats	SDTV	SDTV	HDTV	HDTV
Name	PAL	NTSC	720p	1080i
Picture structure	Interlaced/ progressive	Interlaced/ progressive	Progressive	Interlaced
Pixels per line	720	720	1280	1920
Lines	576	480	720	1080
Total pixels	414,720	345,600	921,000	2,073,600

Tests have shown that 720p is seen in just as much detail as 1080i, in which ultimately only 540 lines are shown per field. This does however depend on content. For still pictures the higher resolution of the 1080i format is preferred by viewers; for moving pictures, 720p usually makes a better effect.

Format	Resolution	Frame rate	Distribution
720p	1280×720	25 fps	Europe, Australia
720p	1280×720	30 fps	USA, Japan, Asia
1080i	1920×1080	25 fps	Europe, Australia
1080i	1920×1080	30 fps	USA, Japan, Asia

How High Definition is perceived

There is no question that a picture recorded using High Definition technology shows more **depth of detail** and brightness – even if it is converted to SD for a standard DVD or for conventional broadcasting using PAL or NTSC, for example.

As well as this, the **16:9 picture format** (→ fig. 6) used for HD, in contrast to the **4:3 side ratio** typical of PAL and NTSC, gives an impression of the image that corresponds better with the human field of vision.

But **1080i** is not necessarily perceived by viewers as the better HD image, even though the resolution is higher than for **720p**. In fact, the full frame process used for 720p can make a more cinematic impression, as film cameras also work with full frames. This might suggest that 1080p, the highest possible full frame resolution, would be an ideal compromise. Far from true: this format would place an enormous strain on broadcasting technology and bandwidth so it is not likely to become widely accepted.

When assessing picture quality, factors such as the source material (sport, fiction etc.), the broadcasting mode (mp2, mp4/H.264 compression etc.) or also the viewing device play a considerable part. This applies particularly when this device (television, projector etc.) has to scale the signal, which is entirely usual for LCD and plasma screens. If the precise resolution of the broadcast signal is not on offer, there can be enormous loss of quality.

6

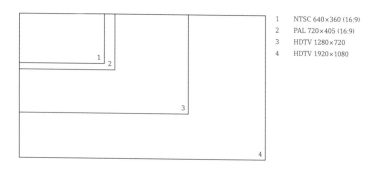

1	NTSC 640×360 (16:9)
2	PAL 720×405 (16:9)
3	HDTV 1280×720
4	HDTV 1920×1080

The advantages of HD productions

- Better quality, detailed picture, more brightness in comparison with SD productions (approx. four times the resolution of SD).
- Suitable for additional broadcasts (stations will be looking for HD content in future).
- Better chance of broader exploitation (HD-DVD/Blue-Ray, HD downloads).

Reproducing HD material

Current flat screens are basically restricted to two technologies: LCD and plasma. The "HD Ready" mark has now established itself. It is awarded only to devices meeting the following criteria:

- Resolution: The device has a resolution of at least 720 lines and 1280 columns in a side ratio of 16:9.
- Input: The device has analogue component input and digital DVI or HDMI input.
- Formats: The device can reproduce 720p (at 25 and 30 frames per second) and 1080i (at 50 and 60 half-frames per second).

– Copy protection: At least one of the digital inputs supports HDCP copy protection.

Different device sizes are available in various resolutions; the broadcast signal is enlarged or reduced appropriately. The technology used for this differs from manufacturer to manufacturer, and is largely responsible for picture quality. Sets that can actually reproduce a 1080 HD signal without changing its size are currently very few and far between, and very expensive.

Tip: *As both LCD and plasma screens usually present full frames and have to convert where applicable for* **interlaced material** *(→ p. 119, 120, interlaced mode), the 720p format is better for display purposes. This is why, for example, German public service broadcasters use this full frame format.*

HD recording formats

There are also many **recording formats** available in the **HD field**. These are **HD Cam** and **DVCProHD** in the professional sector, and **HDV** and **AVC** in the consumer and semi-professional sector. These formats differ primarily, as in the case of SD formats, in terms of the codec used (e.g. 10 bit HD, DVCProHD, MPEG4, MPEG2). A **codec** is the process used to code, decode, compress or expand data or signals digitally. Frequently used codecs are, for example, 10 bit DigiBeta, DVCPro50, DVCProHD, DV, MPEG4, MPEG2, Animation or photo-JPEG.

DVD

The starting-point for a **video-DVD** is **MPEG2** encoded **PAL** or **NTSC picture material**. The sound can be handled either in the **uncompressed PCM format** (stereo), as compressed **AC-3** (stereo, surround) or as a **DTS file** (surround). The programming process – also called **multiplexing** – and creating a master in the form of a **DLT master cartridge**, is known as **authoring**.

For the high definition DVD formats **HD-DVD** and **Blue-Ray**
possible codecs are **H.264** (MPEG4), **MPEG2** and **VC-1**
(WMV). On the audio side they are **MPEG**, **Dolby Digital
Plus**, **DTS-HD**, **MLP** and **LPCM**. Both formats are recommended as successors to the present DVD, but only the
future will show how they are accepted on the market.

Compression

The **bit rate** defines the volume of data for image and sound
information transmitted per time unit. If compression and
resolution remain the same, a higher bit rate will generally
produce higher quality. If different compression processes
are used, the efficiency of the compression process determines the quality along with the bit rate.

The first step towards compressing moving picture material effectively is to reduce the image size and the number of
frames per second (→ p. 119, fps). A more powerful compression codec should not be used until then.

*Tip: Even if the material is only down-scaled to a quarter of
the original image size and 12 fps (rather than 25 fps for PAL),
the data volume has already been reduced to about 12 per
cent – and that is without using a high compression codec.*

Web and computers

There are now countless **codecs** (→ p. 123) available for
moving images on the computer. The most frequently used
are **WMV** (**Windows Media Video**), the HD variant **WMV
HD** and **Quicktime**, though Quicktime is used only as a
superstructure in which different codecs are used, for example Sörenson3, MPEG1 or MPEG4, H.264.

As **Flash plug-ins** are so widespread in **web browsers**,
video integrated into Flash applications is increasingly on
offer (Flash video). The advantage: no other players like, for
example, Windows Media Player or Quicktime Player are
needed.

Software

Editing: Apple Final Cut Pro, Apple Final Cut Express, Avid (e.g. DV Express), Adobe Premiere.

   Motion Design/Compositing: Adobe After Effects, Discreet Combustion, Apple Shake, Apple Motion.

   DVD Authoring: Apple DVD Studio Pro, Adobe Encore, Sonic Scenerist.

   Encoding: Apple Compressor, Sörenson Squeeze, Media Cleaner, Bitvice, MPEncoder, Flip4Mac (Quicktime Player Plug In to create Windows Media Video).

Disciplines and Workflow

It is recommended that the real environment is worked in from the beginning of a digital project, so that the real data are used for each step of the realisation process. This iterative (repetitive) design method is ideal for open systems whose interface constantly changes, regardless of the user, the data offered or the time. Digital procedures are very rarely completed in full. This means that the individual intermediate stages are project-dependent. The stages can be worked through with the aid of **user scenarios**, **use cases prototypes** (paper prototyping), **information architecture** or modelling in **UML**.

Digital procedures, above all for the internet, are frequently directly linked with the client's business models. Examples of this are online banking, online shops but also simple contact enquiries, catalogue orders and complaints management. The systems interlock. E-mails must end up with the right recipient, the car you configure on the internet must match the manufacturer or dealer's production specifications, leasing rates must be calculated in real time. And this means that the designer, too, must have a great deal of technical and economic competence. There will be experts on the individual fields in the team, but the individual designer should still be familiar with the interfaces and problems that can emerge as the project runs its course.

Motion design means creating moving pictures (→ p. 119, Moving Pictures) in the fields of video, film and animation. Unlike narrative films, visual effects are the key here. Examples are creating **titles** and **trailers** for **feature films** or designing **music videos**. The tools used for motion design include After Effects or also Final Cut Pro (→ p. 125, Software).

Interface design is a design discipline dealing with the **user interface** between man and machine. This can mean the design of interfaces for software, i.e. graphical user interfaces (GUI), or also sonic user interfaces (SUI). Conceptual and aesthetic aspects of the interaction are taken into consideration.

Interaction design is similar to interface design, but concentrates less on the actual interface when it is being implemented, and more on the process accompanying the

designed product. Interaction design is based on the particular socio-cultural context and the user's specific multi-
sensual perception (user behaviour). The areas of use extend
from dynamic media content to complex interactive product
systems.

Game design. A game designer devises and designs
computer games. Here the steps include developing characters and the game environment that will be created (level
design). Games are produced for gaming consoles such as
the Nintendo Game Boy, Sony Playstation, MS X-Box etc.,
and also for mobile phones and PCs.

VJ-ing. A **Visual Jockey** (VJ) uses digital mixing technologies from the fields of computers (2D/3D animations) and
video (live cam) to create a visual performance in real time
on a display medium (beamer projection, monitor). In most
cases, VJs use a mixture of existing "scraps".

Mobile application design involves designing for mobile
devices such as **mobile telephones** or **PDAs** (Personal Digital
Assistants). Here, familiarity with current technologies and
standards is just as important as the ability to devise new
ideas and to visualise futures services, or the ability to exploit
a brand's corporate design media-specifically, i.e. appropriately for the screen resolution, colour depth, image rate, data
transmission rates etc. Applications include **information
services**, **mobile soaps** and **games**. A mobile application
designer's tools include Flash Lite, Mobile Processing or the
product manufacturers' individual development environments.

Interactive media environments and installations.
Designing interactive installations and environments includes
working with interactive objects and the way they are presented, and also designing spaces for media experience and
for staging them. Interaction between user and object is built
into this staging and handled creatively in the same way as
the interaction between space and visitors. Examples of
applications here are media experiences of the kind created
for trade shows, at cultural institutions as well as interactive
facades on buildings.

Computational design tackles questions of design by
computer. The design is expressed in the programming
language, as these languages are used to create the design.
Computational design concepts have been around since

1960, but the concept was created largely by the **Aesthetics and Computation** (AEC) Group at the **Massachusetts Institute of Technology** (MIT). The American designer, philosopher and computer scientist Professor **John Maeda** formulated the following principle: a computational designer does not rely on existing tools, but writes his own tools. Computational design's generative systems are formulated by designers. These systems are then run and outputted by computers. The processes that emerge can quite often respond to input, for example, from the user.

Computational design creates processes as well as undertaking classical form and colour design. As many aspects of computational design work emerge only when the product is running, the term generative works is used. The designer defines the rules and states, in other words the system's behaviour, and the system generates the actual work from this. Such a procedure makes it possible to create large quantities of information. For this reason these works are often very complex visually and consist of many autonomous elements, so it is easy to see how closely connected computational design is to presenting information visually.

Computational design tools are the programming languages used to write the work, but also scientific concepts. **Programming languages** include Processing, Java, OpenGL, vvvv, Jitter, Flash, Director and C++.

Web design. Web designers create and maintain websites (→ p. 133). They are responsible for design, construction and the user interface, and for implementing the client's corporate design online (→ ch. Marketing, p. 238). Tools include Adobe GoLive, Macromedia Dreamweaver, Flash and text editors such as BBEdit. Web design is discussed below.

Grid systems for digital and dynamic media

Designers have a number of factors to consider when working with time-based media. Here, aspects relating to perceptual psychology, together with formal and structural aspects, have just as important a part to play as technical elements and the general conditions of the **brief** that these dictate. Even the **output medium** selected can influence the structural and design concept for a piece of media work. So the screen and technical formats have to be considered carefully, and the designer must be familiar with all the general technical, media and design conditions and their particular features.

   Grids are needed in order to place design, media and content elements such as text, sound or video. Dynamic grid systems are created by dividing up an **area**, but also by dividing up **time** and **space**. Certain appearance dates and times relating to dynamic behaviour are allotted to certain elements. Various methods can be used to do this. The various design methods for **dynamic grids** are described below.

   Basic grids. A 2D grid controls conditions within an area, a 3D grid in space and a 4D grid in time. Each grid type also contains elements of these other grids, up to a certain percentage: so two-dimensional behaviour is also fixed on a 3D grid. But one grid type will always be dominant.[3]

   Parametric transfer. One interesting and equally effective way of creating grid systems for animations is the **tracking** principle. The idea is to analyse real and also virtual objects, processes, movements or behaviours and use the values derived as a grid for controlling graphic elements or layers.

   Once parameters have been established for one object, controlling its behaviour, movement, speed, etc., these parameters are transferred to other objects. Here reality serves as a template or pattern in abstract form.

   One easily understandable type of tracking is **motion tracking**. Here, a real movement is recorded and individual segments are "tracked"; the change of position for the tracked segments is captured. These segments or points are used later in the grid to anchor graphic elements placed precisely on the tracked points or segments.

³ Dietzmann, Tanja; Gremmler, Tobias: Grids for the Dynamic Image (2003)

The tracking method can create interesting space-time templates for the behaviour of graphic elements or layers in animations. Information about timing is gained for the elements, as well as positioning in the "space" and "time" dimensions, and this can also be applied to the graphic elements.

Behaviour and sets of rules (library). One important aspect of developing grid systems for moving images is working with self-defined or automated sets of rules. In fact, the characteristics of animated objects are not determined just by their appearance, but also to a crucial extent by their behaviour.

Drawing up sets of rules and modes of behaviour that can then be applied to graphic elements can very rapidly produce more complex animations.

The use of behaviour modes soon turns the designer into a **director**, determining the form and behaviour of his scene and performers. Characteristics such as colour, form, dynamics and the way elements respond to their surroundings can be defined in this way. It is not just about fixing individual **key frames**, for example; in fact the parameters of preset behaviour modes are manipulated for as long as it takes for the objects to start behaving appropriately, in other words with the desired movement and dynamics.

Displacing is a method that is frequently used for designing moving picture sequences. It can influence events in animation extremely powerfully, so is described here as a grid for moving image sequences.

Essentially displacement is a procedure that can be classified as **object tracking**. The values of graphic element qualities such as brightness, colour or tonal value are analysed and transferred to other parameters of the same element or different elements (layers). This makes it possible to create models for the appearance and behaviour of graphic elements very quickly.

Here, the model or motif whose values are picked up and transferred is usually not a visual component of the animation, but a **displacement map** created expressly for this purpose. In principle it is a matter of using an object's qualities and transferring them to other objects. At the same time, the changes of value undergone by a particular element or layer within that particular time-line are also transferred.

This procedure is often used in post-production. But if it is applied more universally, and used as a model for controlling the graphic elements of the animation as a whole, then it is a method that is as powerful as it is simple for generating grids swiftly and efficiently. As well as the selected parameters transferred, a large number of other parameters are available, and these too can be applied to the values derived.

The visual appearance of particles and graphic elements can be controlled in this way. These values can also determine behaviour in space and time of the elements used within the sequence.

Interactive movement. Grids for interactive systems are based on the pixel. In addition and as a rule, classical 2D grids are used that model the programming and supply descriptions for time and interactive sequences.

When designing interactive movements, the key issue is not to create a fixed framework in which all the elements are then placed, but to design a **dynamic structure**. This dynamic structure presents the contexts of and links between the content and functional aspects visually, thus making them available to users as an **interactive structure** – a **navigable structure**.

If interactive systems are seen as interactive, dynamic structures, then both the structure designed and its behaviour can be used as a grid, i.e. as a set of rules for the **interface** (→ p. 126). As in complex animations, the rules governing an interface are explained by its structure, the definition of behaviour and the way all parts interact.

The **interaction** component is then added, which should define system behaviour modes for all of the user's actions. These guidelines should have an inherent consistency that makes it possible for the user to understand the system quickly and then be able to navigate independently and with a clear view ahead.

Motion graphics

Usually, the aesthetics and overall appearance of an **animation** derive from the interplay of various factors. Several grids are used for one animation, with one grid usually dominating.

Very often, key frames are placed by hand so that individual details can be manipulated. An **effect** is defined through a sequence of several key frames, defining the choice of speed and form of movement between the key frames along a time axis. **Key frames** allot a very particular quality (movement or position) to a very definite moment (e.g. coordinate).

Some animations manage without grids. They are built up on manually placed key frames only. Here, links are created between **layers** and effects, but relate mainly to an individual layer and do not follow previously defined statements, as in a grid.

Hint: This approach does provide unlimited creative scope, but considerably more time and effort are needed to create a movement or dramatic sequence. Moreover, sometimes it is not possible to avoid animating manually. Designing motion graphics in a corporate design context often requires precise manual construction of dynamics and a dramatic line, as well as the use of grids.

Web Design

Designing and programming **websites** is an aspect of digital media production. A whole project on the **World Wide Web** (→ p. 115), usually consisting of several documents, files or resources brought together and linked by uniform navigation, is called a website. Internet services extending beyond the web are not part of a website, but can be part of the internet presence of which the website is part.

The terms "site" and "page" are often confused in this context. The "site" defines the location as a whole, rather than a particular "page" within it. A **website** always means the entirety, and is placed on a server (host) on the World Wide Web, while what appears in the browser is an **individual web page**. The starter page in a website is called the **homepage** – or also the gateway or portal.

The **role of a website** is to create **presence, communication** and **transparency**. Maintaining a website mainly involves attending to structure, layout, functionality and **content**, where appropriate using a **Content Management System** (CMS). Different demands can be made on the quality of a website, but there are important aspects such as the quality of the content, of the information and of intuitive user guidance that should always be guaranteed.

Good web design is based on the following rules:

- Simple, clear and rapid information presentation.
- Reduction to essentials.
- Logical and manageable navigation.
- Uniform design principle applied throughout.
- User-friendly download times and correct display in all major web browsers.

Page size. When organising a design area, care should be taken that the size of a web page matches monitor resolution, so that important elements are not left outside the visible realm. For a long time, the optimisation for a resolution of 800×600 pixels was considered ideal, but now a resolution of 1024×768 pixels has been adopted.

However, this area cannot be fully exploited when design-ing a website, as the monitor does not just display the page as such, but also elements from the browser and operating system such as menu lists and scrolling bars. These vary in size according to browser, operating system and the user's own settings.

Tip: *A design area of 971×608 pixels is considered "safe" for a page that is to be optimised for 1024×768 (figs. 7, 8).*

Typefaces can be displayed only if they are also installed on the computer using the site. This means that web designers can use only the fonts installed as standard in their operating systems (→ p. 140, fonts), like for example Arial and Verdana. Times New Roman and Courier are not recommended for websites (→ ch. Typography, pp. 93 – 95, Fonts on a display screen). If other typefaces are to be used (e.g. a house font in the menu), this can be done only by placing appropriate image data.

Note: To guarantee barrier-free access (→ p. 142), the cor-responding text must be set up along with these graphics. Using images also means that the page download time is increased, and a new image has to be created each time corrections are needed.

Not all browsers display the same fonts in the same way, as anti-aliasing (→ ch. Typography, p. 94) differs from font to font. Some operating systems, particularly older ones, do not use anti-aliasing at all.

Every **image or graphics file** should also be as small as possible. The number of graphics per page should be con-sidered very carefully. Even if the file size for a photograph is reduced to 15 KB, the page size increases to over 150 KB when a website uses ten pictures. This can badly affect down-load times.

Navigation elements are used to ensure that the content is arranged clearly and its aims well defined. As well as creating movement, by means of the leads from item to item, they should above all help users to find their bearings. Users should always know exactly where they are in the system, where they have come from, how to get back there, where they can navigate to and where certain information relevant to them is to be found.

Buttons are graphically prepared switching areas that trigger an action when clicked on (→ ch. Marketing, p. 230).

What is needed to set up a website?

1. An image processing program to create the screen design and process the graphics (e.g. Adobe Photoshop).
2. Web space (via a provider).
3. An **editor** to write the source code (text editors or WYSIWYG editors, → p. 136).
4. Several **web browsers** to test correct display.
5. An **FTP program** to upload the code data to the web space.

Programming / The source code

Content and design should be kept separate, in order to create websites that are easy to maintain, with a minimum of code.

(X)HTML is used to structure content. **CSS** (Cascading Style Sheets) (→ p. 139) are used to define the layout of the subject matter, such as positioning on the page, colours, fonts, type sizes etc. These stylistic requirements should not be written into the HTML document directly, but should be linked in through external CSS files.

HTML

HTML stands for **Hypertext Markup Language**. This language is based on a hierarchical structure (→ pp. 131, 144, 145). HTML describes the structure and qualities of documents. **Hypertext** stands for the structure and composition of documents on the internet and does not use a linear approach. **(Hyper)links** (cross-references) can be made to other texts. These jumps are leads or links to other documents. They are executed by **mouse clicks** on fixed textual passages, images or graphics.

HTML is not dependent on any particular platform as a language, and is not tied to any specific operating system or browser.

Several standard web browsers are needed for testing and displaying the final result when creating HTML documents, and so is editor software for inputting the HTML codes. Theoretically even the simplest **text editors** are suitable for this, but it is recommended to use special HTML editors. These include features like highlighting the various HTML commands in colour, and they will often automatically complete and validate the code.

So-called WYSIWYG editors (What You See Is What You Get) are visual editors offering a preview of the page and showing the code only if wished. Many developers are sceptical about WSIWYG editors, as they do not usually produce suitably spare, standardised code.

HTML **commands** are implemented by so-called **tags**. A tag consists of two brackets, "<" and ">", with the command between them. As a rule, tags consist of a start and an end tag. The **end tag** always includes a slash ("/") after the introductory "<" character. Example: start tag: <p>; end tag: </p>. This tag encloses a paragraph ("**p**" stands for **paragraph**). There are also single tags, for example <br>, which introduces a line break.

Tip: *in* **XHTML** *(→ p. 139) all the tags are closed, for example the line break tag looks like this: <br/>.*

Examples of other HTML tags

Headings. HTML prescribes six standard headings. Each heading is automatically a paragraph in its own right. Standard headings are introduced by the following tags:
<h1> </h1> encloses a heading of the first order.
<h6> </h6> encloses a heading of the sixth order.
Emphases are defined by <em> </em>.
Include graphic: <img src="image1.jpg">
Create link: <a href= linktarget.html >link text</a>

All the qualities affecting the way the content looks, in other words the design, should be defined in a CSS file, not in HTML code (→ p. 139).

As CSS was formerly not adequately supported by all browsers, and also because the first CSS version did not offer good element positioning opportunities, attempts were frequently made to layout using HTML tags.

So, for example, tables (<table> </table) were frequently used to fix column widths or to allot particular positions to certain elements. Nowadays <table> should be used only for actual table material.

The <font> tag, formerly often used to define font, type size and colour should not be used today, as all of these qualities can be defined by CSS, which now covers almost all design possibilities.

Some examples of special HTML characters

A-Umlaut (Ä)	&Auml;
a-Umlaut (ä)	&auml;
Paragraph sign (§)	&sect;
Copyright sign (©)	&copy;
Square sign (^{2})	&sup2;
Pound sign (£)	&pound;
Dollar sign ($)	&dollar;
Euro sign (€)	&euro;
Yen sign (¥)	&yen;

Basic framework of an HTML file

Every HTML document has a basic framework that sends commands to the browsers to enable them to display the document. The following summary shows how an HTML document is made up.

```
<html>
  <head>
    meta http-equiv="content-type" content="text/
    html;charset=iso-8859-1">
    <title>
    </title>
  </head>
  <body>
  </body>
</html>
```

Every HTML document starts with <html> and ends with </html>. This tag tells the browser that this is an HTML document.

The head

The first tags to appear between the <html> </html> tags are the tags for the head: <head> and </head>. Between the head tags are the tags for the page heading: <title> and </title>. The information between the title tags defines the text shown in the browser's title bar. Information for search engines can also appear between the head tags (→ ch. Marketing, pp. 228, 229, search engines).

The instruction <Meta http-equiv="content-type" content="text/html;charset=iso-8859-1"> tells the browser what type of content the document contains: HTML text, and the character coding the HTML file is using, in this case the ISO-8859-1 character set. This contains the standard Western character set. This information is important, as otherwise there is no guarantee that the browser will display the characters on the website as required.

The body

Next come the tags for the actual content of the HTML page,
the body: <body> and </body>. The tags found between
the body tags define the content and the way in which it is
structured.

XHTML

XHTML is being promoted as the successor of HTML.
HTML is based on **SGML (Standard Generalized Markup
Language)**, whereas XHTML is based on **XML (Extensible
Markup Language)**.
    XML is a metalanguage defining any number of markup
languages. These languages can describe text documents,
vector graphics, multimedia presentations, data bases or
other kinds of structured data.
    XML was developed by the **World Wide Web Consortium**
(W3C) (→ p. 146, Tips and Links) and declared the standard.
XML is a further development of, and complements, the classical metalanguage SGML.
    Unlike HTML, XHTML is case sensitive, in other words it
distinguishes between upper and lowercase characters.

Tip: *The lowercase characters prescribed for XHTML should
be used for all element and attribute names in HTML as well
so that the pages will be compatible in future.*

Tip: *The W3 Consortium (W3C) fixes XHTML and CSS standards (→ p. 146, Tips and Links).*

CSS

The abbreviation CSS stands for Cascading Style Sheets.
Style sheets are style templates. CSS is used to define the
formatting and positioning of HTML elements, including setting type, line spacing, fonts, colours, element positions and
many more. CSS instructions should not be placed directly in
the HTML file, but in an external file that is then integrated
into the HTML file. This ensures that content and layout are
separated, that the style requirements can be handled and

maintained with ease, and it also reduces the size of the pages, as the CSS file has to be loaded only once.

Fundamentally it is possible to choose **fonts** (→ ch. Typography, p. 90) for websites that are installed in operating systems as standard fonts or also as system fonts.

The operating systems to be considered here are Mac OS, Windows and Linux. As the same standard fonts are not available on all operating systems, alternatives can be given for a particular font, e.g. font family: Verdana, Geneva, Arial, Helvetica, sans-serif.

If the font mentioned first is not available on the user's operating system, the next is used automatically. If none of the named fonts is installed, it makes sense to give a generic name last, so that the browser can at least choose a similar typeface: serif = a font with serifs, sans-serif = a font without serifs.

Tip: The W3 Consortium recommends putting fonts with spaces in their names in quotation marks.

Colours are usually given in **hexadecimal** form (→ ch. Design, p. 14) and have to be translated into the hexadecimal system. The hexadecimal value consists of six places. The first two places are for **red**, the middle two for **green** and the last two for **blue**; they are always preceded by the hash sign ("#"). Example: #00ccff. (For colours with six identical places, it is permissible to use just three, in other words #000 instead of #000000.)

It is also possible to define colours by their colour names, though here only the 16 defined as standard by the W3 Consortium are available. An example: lime, which corresponds to the hexadecimal value #00ff00.

Other programming languages

JavaScript is a programming language in the style of C/C++ to complement HTML. It is possible to build special functions (e.g. tickers) into websites with Javascript. Like HTML, JavaScript is platform-independent, and usually works with any browser.

Java is also a platform-independent programming language in which programs do not have to be rewritten for each operating system. Unlike JavaScript, it is not the source code that is transmitted here, but a precompiled program code. This means that users can see what the commands in JavaScript are, as they are written directly into the HTML file. In the case of Java itself, the user has no way of seeing what the command code is, and thus it is also not possible to work out the code or the source text for the program.

PHP Hypertext Processor (PHP) and **Active Server Page (ASP)** are programming languages whose programs are executed directly on the server, not in the user's browser. Unlike JavaScript, they are used for processing data on the server.

Barrier-free websites

The term **barrier-free** on the web means using and keeping to prescribed standards when programming websites (W3C, WAI, BITV). These standards are intended to make the internet a medium that is accessible to all users. If a website deviates from these standards, then it is not just the physically disadvantaged who encounter obstacles. The aim of a barrier-free website is to display the content offered in such a way that it can be accessed without difficulty and independently of the output device (e.g. web browser, PDA device, language output program, mobile phone or other mobile devices); this is also called **accessibility**. A barrier-free website is displayed in the best possible way independently of the chosen presentation or output form.

The key requirement for barrier-free internet pages is keeping to web standards (valid HTML/XHTML). The strict separation required between content (text, images etc.) and

layout is achieved by the correct use of Cascading Style
Sheets (CSS). No design compromises are necessary.

Tip: *The W3C's Web Accessibility Initiative (WAI) provides
international guidelines for barrier-free websites in their Web
Content Accessibility Guidelines (WCAG). At the time of writing, a website can be tested for validity of these guidelines free
of charge online on the W3C site (→ p. 146, Tips and Links).*

Basic rules for barrier-free access

1. Construction using valid (X)HTML.
2. Strict separation of content and layout.
3. HTML used exclusively for arranging and structuring
   the contents of a page.
4. CSS is used to fix the page design.
5. The homepage is not cluttered, the number of functions
   is manageable.
6. Appropriate text-based alternatives are offered for non-
   text-based content.
7. Navigation is not made dependent on the input device.
8. Tables are only used to reproduce tabular content, not to
   fix a particular layout.

Usability

An effective navigation system is essential: whether for software or a website, it has to be clear how the interface can be
used actively. The process behind this should be considered
step by step, and the best possible user-friendliness offered.
If users do not find what they are looking for on the internet,
and can also not find their bearings, because, for example,
countless (unnecessary) animated ads slow down the search
for a product or make it difficult, then they will leave the site
and look elsewhere. Jakob Nielsen[4] said in this context:
Usability rules the web. Basically, if the customer can't find
a product, then he or she will not buy it. Key factors for
usability, in other words for a high level of user-friendliness
include efficiency, effectiveness and satisfaction.

⁴ Born 1957 in Copenhagen; IT expert, writer and advisor on software and web design.

In order to create a **structure** for a **graphic user interface**, the following questions and criteria should be addressed:

1. Simply presented information.
2. Content reduced to essentials.
3. Manageable, logical structure.
4. Intuitive usability.
5. Uniform appearance, consistently designed.
6. Good design and appealing graphics.
7. Quick delivery of information.

Tip: *Anyone wanting to design a particularly user-friendly website should offer a sitemap (table of contents) of every page, and a search field on the homepage.*

A **sitemap** features a complete, hierarchically structured listing of each individual document (page) in an internet presence or a website.

Presentation and Structures

Graphic user interfaces can be structured in various ways. The following overview shows examples of the most important structures and their characteristics.

Linear structure (fig. 9)

1. Linear screen arrangement.
2. No active user intervention possible.
3. Each item of information built on others.

Jump-line structure (fig. 10)

1. Linear screen arrangement.
2. User can reach any page from the homepage.
3. Little interactivity.

Tree structure (fig. 11)

The classic structure is the tree structure. Users reach various subsidiary pages from the homepage.

1. The branches offer various navigation possibilities.
2. Logical structure (hierarchical structure).
3. Suitable for many applications like for example information and learning schemes and websites.

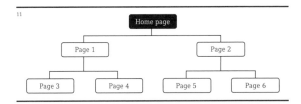

Net structure (fig. 12)

1. No clear hierarchy.
2. Smaller, interlinked subsidiary systems.
3. Nodes correspond with the main menu page.

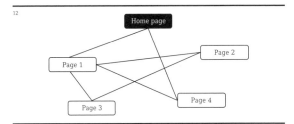

Single frame structure (fig. 13)

1. Experienced as a single page by users.
2. No hierarchies.

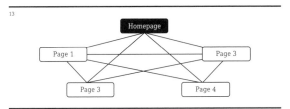

Tips and Links

Digital design and media production

Adobe Creative Team: *Adobe After Effects 7.0 Classroom in a Book.* Adobe Press, Berkeley (2006)

Taylor, Jim: *DVD Demystified Third Edition.* McGraw-Hill/TAB Electronics, New York (2006)

Videomaker: *Guide to Digital Video and DVD Production.* 3rd edition. Focal Press, Burlington (2004)

Perception and psychology

Boff, K. R.; Kaufman, L.; et al., (Eds.): *Handbook of Perception and Human Performance.* John Wiley & Sons, New York (1986)

Gibson, J.J.: *The Senses Considered as Perceptual Systems.* Greenwood Press, Westport (1983)

Goldstein, E. Bruce: *Blackwell Handbook of Perception.* Blackwell Publishing, Malden (2001)

Web design, internet, programming

Browser statistics:
www.w3schools.com/browsers

Free Photoshop browser templates:
www.webdesignerstoolkit.com

Internet Corporation for Assigned Names and Numbers (ICANN): www.icann.org

Tim O'Reilly on Web 2.0:
www.oreillynet.com/pub/a/oreilly/tim/news/2005/09/30/what-is-web-20.html

World Wide Web Consortium (W3C):
www.w3.org

W3C's Web Accessibility Initiative (WAI):
www.w3.org/WAI

Networks and associations

European Multimedia Associations Convention (EMMAC): www.emmac.org

6.0

It's hard to imagine today, but before the invention of movable letters – back in the days when words were still recorded on clay tablets, papyrus or parchment – all books and documents had to be copied by hand. This was an incredibly laborious and lengthy process.

In printing, a lot has changed since the woodcut, one of the oldest printing techniques known to mankind. Nowadays, designers can choose from a plethora of printing options – whether letterpress, offset or gravure; whether computer-to-plate or direct imaging, or whether standard or spot colours – when the deciding just how to commit their ideas to paper. Despite the rapid advance of electronic media, this versatile, cellulose-based medium remains the central element of our visual communication.

To make the most of the latest technological advances and create just the right framework for your own work, we invite you to join us in an extensive exploration of the different paper types, applications and finishing methods, formats, grammages, inks and printing techniques that will shape the final product and determine its look, feel and intended impact.

Expert bookbinders, printers and publishers round out the chapter with insider tips on a range of topics from trimming and creasing choices, grid types and pre-flight checks, to layout, editing and colour management to give your work that extra edge, and help you avoid any pricey mistakes before your masterpiece goes to press. Happy printing!

Production

6.1	**Paper Formats**	152
6.2	**Paper**	156
	Paper types	161
6.3	**Reproduction**	167
	Colour management	169
	Halftone screens	172
	Artwork and control	176
	Print forms	183
6.4	**Printing**	186
	Printing process control	187
	Printing processes	189
6.5	**Post-press**	197
	Surface finishing	197
	Folding	199
	Trimming	200
	Binding processes	201
6.6	**Production in Commercial Practice**	203
6.7	**Tips and Links**	207

Paper Formats

In Germany, the standard paper format sizes are stipulated by the Deutsches Institut für Normung (DIN, German Institute for Standardization) in standard **DIN 476**. The German standard is used as the basis for the corresponding international standard **DIN EN ISO 216** of the International Organization for Standardization (ISO). The A series (preferred series) and the B series in the ISO/DIN standard have been accepted by almost all countries. Usually, the only differences are the permitted tolerances. But some countries still retain traditional methods, most of them less systematic and non-metric. For example, the most common format in the USA and Canada is 8.5 × 11 inches (→ p. 155, letter). The B series is also slightly different in Japan: B4 = 257 × 364 mm.

The ISO/DIN series have considerable advantages over non-standard formats. The reference format for the A series is AO, with an area of a square metre (1 m²). The shorter side of the sheet is in a ratio of 1 to root 2 (1.414…). Each format is twice or half as large as the next format in the series. The next smaller format is always reached by halving the long side of the initial format. The format classes "0" to "8" show how often the A0 starting format has been halved. The A and B series are usually final formats. The divisions of an A sheet are shown in the following overview.

The **ISO/DIN series** are sized in millimetres. The tolerance is ± 1.5 mm up to 150 mm, ± 2 mm for sizes up to 600 mm, and ± 3 mm above that. The oversize 2A0 and 4A0 formats exist only in the DIN standard, not the ISO standard.

ISO/DIN A series in mm		B series in mm		C series in mm		D series in mm		E series in mm	
4A0	1682×2378								
2A0	1189×1682								
A0	841×1189	B0	1000×1414	C0	917×1297	D0	771×1091	E0	800×1120
A1	594×841	B1	707×1000	C1	648×917	D1	545×771	E1	560×800
A2	420×594	B2	500×707	C2	458×648	D2	385×545	E2	400×560
A3	297×420	B3	353×500	C3	324×458	D3	272×385	E3	280×400
A4	**210×297**	B4	250×353	C4	229×324	D4	192×272	E4	200×280
A5	148×210	B5	176×250	C5	162×229	D5	136×192	E5	140×200
A6	105×148	B6	125×176	C6	114×162	D6	96×136	E6	100×14
A7	74×105	B7	88×125	C7	81×114	D7	68×96	E7	70×10
A8	52×74	B8	62×88	C8	57×81	D8		E8	
A9	37×52	B9	44×62	C9	40×57	D9		E9	
A10	26×37	B10	31×44	C10	28×40	D10		E10	

The strip formats from the A series, e.g. 1/3 A4 = DIN long (DL): 99×210 mm, are derived by division.

US paper formats do not follow a uniform pattern and were originally based on the inch (unit: in). The conversion values in millimetres (mm) are rounded up or down appropriately. The following overview shows the most common US and Canadian formats.

Name formats	US ANSI formats	inch	mm	CAN formats	mm
				P6	107×140
Invoice		5.5×8.5	140×216	P5	140×215
Executive		7.25×10.5	184×267		
Legal		8.5×14	216×356		
Letter (US-Standard Letter)	A	8.5×11	216×279	P4	215×280
Tabloid (Ledger)	B	11×17	279×432	P3	280×430
Broadsheet	C	17×22	432×559	P2	430×560
	D	22×34	559×864	P1	560×860
	E	34×44	864×1118	E	34×44

There are also **untrimmed** or **oversize formats**. These are larger, as they are not trimmed to an A series format until after printing, folding and binding. The following overview shows the ISO/DIN RA (untrimmed sheets) series.

ISO/DIN RA series	in mm
RA-0	860×1220
RA-1	610×860
RA-2	430×610
RA-3	305×430
RA-4	215×305

Paper

In past cultures, stone, metal, wood, wax or clay tablets were used to convey and record information. These materials were gradually replaced by more flexible ones. **Papyrus** can be shown to have existed as a writing material in Egypt since the early 3rd millennium BC. This writing material was made from the papyrus bush[1] – a reed-like marsh plant. The stems of the bush were cut into strips, laid on top of each other crosswise and then pressed, hammered, smoothed out and dried.

Recent findings show that paper must have existed in China as early as 60 BC.

Paper forms

Paper can be supplied by the manufacturer or the wholesale trade in **reel** (fig. 1) or **sheet form**. **Formatted paper** (fig. 2) is cut to the size required in the paper factory and then packed appropriately, usually in a size optimised for a particular print format. This reduces **cutting waste** at a later stage. As paper prices are usually calculated by weight, this can reduce paper costs.

Paper manufacture and qualities

Paper has different qualities according to the way it is manufactured and according to its composition. The basic raw material for paper is **pulp**. Pulp is a material made up of cellulose fibres derived chemically or mechanically from vegetable raw materials, particularly wood. **Cellulose** is a chain molecule built up from dextrose units (polysaccharides). As the main component of the plant material, cellulose gives the plant its mechanical stability.

Wood-containing papers contain over five per cent of mechanical fibres derived from wood (woodpulp). As **woodfree papers** are papers with at the most five per cent of their weight made up of mechanically lignified fibres, the term woodfree is not essentially correct. **Rag-containing papers** contain at least ten per cent rag and are used mainly for banknotes and documents.

---

[1] Latin: Cyperus papyrus

Sheet making

In industrial paper production, the sheet making process takes place on a paper machine. The **paper pulp**, also called **paper stock**, is 99 per cent water. After being cleaned several times, it runs from the **stock vat** into the **sieve section** of the **Fourdrinier wire paper machine**. A large proportion of the water runs off on the drainage sieve. This creates the paper structure.

Fillers are additives in paper manufacture, and fill in the gaps between the **paper fibres**. As a rule, fillers are minerals such as China clay, also known as **kaolin**, or **calcium carbonate**. The filler content of the finished paper can be up to 35 per cent.

The composition of the fillers determines the opacity, the degree of whiteness and also the smoothness of a paper. The **opacity** (lack of transparency) identifies the amount of light that can pass through the paper. Paper that is to be printed on both sides should have a high degree of opacity. A higher proportion of wood in the paper and fillers such as China clay, talcum or titanium dioxide increase opacity. Ink resistance is achieved through the addition of a glutinous substance known as **size**.

The role of size is to guarantee that the paper particles will stick together, to make it possible to write on the paper.

Wet beating (high level of beating) or **free beating** (low level of beating) of the paper fibre influences the paper's resistance. Thus, for example, highly sized paper is very smooth and tear-resistant. Fig. 3 shows an ink line on heavily sized paper in contrast with unsized paper, as in fig. 4. The size is usually added to the paper mass before it is processed. For special kinds of paper, the surface is sized during the drying process in the paper machines.

Various quantities of **dye** are also added to the paper pulp; these include **optical brighteners**.

Marks on the paper brought about by different paper thicknesses are called **watermarks**. So-called **genuine watermarks** are created by compressing (light watermark) or enriching (shadow watermark) the mass of paper fibres in the sieve section of the paper machine. The process is carried out with a watermark roller (watermarking dandy).

3

4

Semi-genuine watermarks are stamped into the paper after it has left the sieve. **Non-genuine watermarks** are made outside the paper machine by printing with a colourless varnish (→ pp. 197, 198) or by stamping (→ pp. 198, 199).

Basic machined papers are ready at the end of the manufacturing process without finishing (→ pp. 197, 198, 202). But an even smoother surface can be created by satinising, for example.

Paper finishing

Paper is usually finished after the initial manufacturing stages. One form of **finishing** is **coating** in a **coating machine**. Many papers are coated to improve their surface smoothness, gloss, whiteness and printability. The **coating colour** used for this consists of natural pigments, a binder and various additives.

Calendering – also called **satinising** – is another form of paper finishing. Smoothing the surface in the **calender** makes the surface glossier, the so-called ironing effect. Calenders themselves are special machines for calendaring or **satinising** paper. Essentially, they consist of a system of cast steel and paper rollers. These are placed one on top of the other and the paper passes through them in a snake configuration. Here the polished steel rollers perform the actual smoothing function. There are also other finishing processes. For example paper can be **stamped** (e.g. structural stamping). This increases their surface area, and makes them longer-lasting and more robust.

The quality of the **paper surface** is very important if paper is intended for printing; it also very stimulating in visual or tactile terms for users. Other paper characteristics are also important for some printing processes (→ pp. 189–196). Thus, for example, papers intended for web printing on rotary presses must be particularly tear-resistant, and letterpress papers very elastic, so that they can adapt to the pressure of the plate. Colour printing requires high **dimensional stability** – a measurement that indicates the extent to which paper dimensions change with the moisture content. The paper must not stretch in printing, as the colours cannot otherwise be printed on top of each other precisely.

The **picking resistance** of paper indicates how much force
has to be exerted by vertical tension to detach particles from
the surface. Picking resistance is particularly important in
offset printing (→ pp. 189–191).

The **ash content** of a paper identifies the proportion of
inorganic substances that remain behind as ash when the
paper is burned.

Grain

The direction – also **grain** – of the paper is the lie of the fibres
in the paper, which derives from the manufacturing process.
This means that the paper has different qualities relating to
two directions. When paper is machine made, the grain is the
direction in which the paper runs through the machine. The
fibres lie in a particular direction in relation to the flow (fig.5)
of the wet paper pulp (→ p. 157).

Golden rule: In the case of books or brochures with
several pages, the grain of the paper should run parallel
(fig. 6) with the gutter (→ ch. Typography, p.82, fig.48).
Otherwise, the moisture in the size would cause the paper
to swell and become wavy at the gutter. That would not just
make the printed book look unappealing; it would also affect
its handling, when opened, for example (**opening characteristics**), or when the pages are turned. If the paper is folded
(→ pp. 199–200) or creased (→ p. 200) against the grain, the
fibres in the paper break.

Sheets can be cut in **narrow web** (also long grain) or
wide web (also short grain) from the width of a paper web.
In the case of narrow web, the paper fibres run parallel with
the long side of the sheet; for broad web, they run parallel
with the shorter, narrow side (fig. 7).

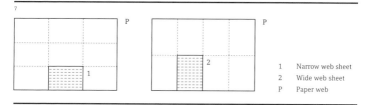

1   Narrow web sheet
2   Wide web sheet
P   Paper web

Golden rule: Always specify first the width, then the height when giving dimensions. Examples: 14.3×21.4 cm = **portrait** format; 21.4×14.3 cm = **landscape** format. To avoid misunderstandings, it is also customary to add the terms landscape or portrait in addition to the dimensions.

There are various **test methods** that can be used to determine the grain of a paper:

The **tearing test:** A piece of paper is torn lengthways and widthways. The paper will tear more easily and in straighter lines parallel with the grain (figs. 8, 9).

The **fingernail test:** Both sides of a sheet of paper are drawn through the thumb and the middle fingernail. If the edge of paper remains smooth, it is parallel with the grain; it appears wavy across the grain.

The **moisture test:** The paper is moistened along its length and width. Parallel with the grain the paper rolls up, across the grain it appears wavy.

The **bending test:** The paper is bent. The paper offers less resistance parallel with the grain. The advantage of this testing method is that the paper is not damaged.

Paper weight

The weight of paper is measured per unit of area (or **area weight**), in grams per square metre (g/m²). **Paper grammages** range from seven to about 225 grams per square metre. The commonest paper and card weights are summarised below.

Lightweight printing papers	approx. 40 g/m²
Newsprint	approx. 50 g/m²
Poster papers	approx. 60 g/m²
Art papers	90 to 135 g/m²

It is not unusual for the weight of paper to be erroneously equated with paper thickness. In fact, papers of the same weight can have different thicknesses. The relationship between paper thickness and paper weight is defined as **volume**. Here a distinction is made between one, one-and-a-half, two and two-and-a-half times volume.

Paper types	Volume
Glossy illustration paper	0.75 – 0.8
Matt coated illustration paper	0.75 – 1.0
Slightly voluminous paper types	from 1.0 (1.1.–1.3)

Tip: *Block thicknesses in book or brochure production cannot be calculated accurately. Accurate figures can be ascertained by making a dummy. It is recommended that paper from the same batch as the production run is used for a dummy, as paper thicknesses can vary in manufacture. This means that the thickness of a specimen sheet is different from the paper used for a print run, and this can lead to different spine thicknesses.*

A **dummy** is intended to demonstrate the technical characteristics of a product to be printed such as format, size, paper quality, finish and binding. What is missing is the actual printing process: the pages are left blank.

Tip: *A dummy is used to ascertain the thickness of a book block. Bookbinders also use a dummy to establish a dimension sheet or front view (fig. 10). The front view is used to establish details for the cover or the binding. It will lay down the card width and height in the case of hardbacks.*

Paper types

Paper types are characterised in various ways. Possible criteria here are the manufacturing method, composition, weight, surface qualities or the purpose for which they are used.

Coated and uncoated papers

Coated papers are paper types that have been coated with a substance containing a pigment on one or both sides. Very thin coated papers are also known as LWC papers (Light Weight Coated). **Uncoated papers** have no coating. Coated papers can also be broken down into other categories.

Art papers, for example, are high-quality papers coated elaborately on both sides for prestigious, colour printing

products. They usually have an even and smooth surface.
These papers are available with matt, semi-matt and glossy
qualities. Pictures of the highest halftone grades can be
reproduced in letterpress or offset. **Illustration printing
papers** are coated, often also calendered, and available in
various qualities.

Tip: *Illustration-quality papers are often used to replace the
more expensive art papers.*

Cast-coated papers and cards are especially glossy. However, this gloss is not produced by calendering, ($\rightarrow$ p. 158),
but by rolling the damp or dampened printing substrate with
a highly polished chromium-plated drying cylinder.

Board and cardboard

Cardboard is a flat material with largely uniform fibrous layers (from 225 g/m²). While board always has a finished surface
and consists of high-quality materials, cardboard is made of
lower-quality raw materials. **Board** has a grammage between
paper and cardboard, and is made of the same raw materials
and additives as paper. A distinction is made between single-
layer and multi-layer boards and cardboards.

　　Board and cardboard cannot be precisely defined by
weight. When working with cardboard and board, attention
must also be paid to the grain ($\rightarrow$ pp. 159, 160).

　　A distinction is made between various kinds of cardboard. So-called **pasteboard** is used most frequently. The
various thicknesses of pasteboard or **machine pasteboard**
are given in millimetres, e.g. 2.0 mm. The various weights of
hardboard are given in kilogrammes.

Tip: *Hardboard has a solid, dense surface and is particularly
suitable for large areas of embossing on book covers.*

Bristol board is a kind of board made up of three or more
layers; the outer layers are woodfree, but the inner material
can contain wood. Bristol board is uncoated. It is robust and
well suited to offset printing and further processing. Typical
uses are for postcards, envelopes or packaging.

Chromium sulphate board is a board that is usually coated on one side only, with a weight of about $18\,g/m^2$. The primary product is chromium substitute card, smooth on one side, with a light-coloured middle layer with a high wood content and a covering layer that is woodfree on one or both sides. It is also known as folding boxboard. **Chromolux board** is a brand name for a glossy cast-coated board that is white on one side.

Duplex board is a multi-layered board with a grey middle layer, grey back and a woodfree covering layer, or one containing very little wood.

Special papers

Parchment/greaseproof paper is a boil-resistant, fat- and waterproof paper with an impermeable surface. The initial product for parchment paper is an absorbent base paper that is treated in a sulphuric acid bath in order to seal the surfaces. This paper is used above all for packaging fatty products (e.g. margarine). **Pergamin** is made of finely beaten pulp and is a largely greaseproof but not moisture-proof paper type that is heavily calendered and thus looks relatively transparent.

Tip: *Parchment papers react particularly strongly to size in further processing and often look corrugated. They also take a very long time to dry after printing.*

Carbon-copy paper is a thin (30 to $40\,g/m^2$), machine-finished typing paper, usually woodfree and well sized.

Carbon papers consist of a substrate with carbon ink that conveys the impression through to the next sheet when printing.

WP papers are papers made of on hundred per cent waste paper.

Fine papers are usually the **best quality papers**. Particular attention is paid to the firmness of the surface in manufacture, to good and even opacity and to very high printing quality. Fine papers often have watermarks. **Medium-fine papers** have over five per cent of mechanically recovered wood fibres (mechanical pulp).

High gloss papers are cast-coated on one side and not calendered (→ p. 158).

Chromium papers are wood-containing or woodfree papers coated on one side. The coating is always watertight, and is formulated with a view to making the paper highly suitable for offset printing (→ pp. 189–191), as well as for stamping, varnishing and bronzing. Chromium paper is used mainly for labels.

Coloured papers are tinted, varnished and patterned papers, also velour, bronzed or marbled papers.

Vat papers are handmade papers drawn "from the vat" with a sieve with a typically uneven edge. Machine-made imitations of this kind of paper are also available today.

Japan papers – also called "washi" (wa = Japan and shi = paper) – are made of native Japanese plants such as kozo, mitsumata, gampi and kuwakawa.

Papers by use

Bulky papers – also called voluminous papers – are soft, elastic papers. Thin papers – also called India or Bible[2] papers – are **low grammage papers** made of rags and bleached pulp and with great strength.

Offset papers are woodfree and wood-containing un-coated papers (→ pp. 161, 162) as well as uncoated recycled papers in calendered or machine finishes (→ p. 158). They are especially well suited for use in offset printing.

Magazine papers – also art papers – are uncoated, usually wood-containing, calendered papers with a high proportion of filler that are particularly good for reproducing pictures. These papers are used above all for gravure-printed magazines.

Laser printing papers have an even, specially prepared surface to ensure the best possible toner adhesion and to support the instant electrical discharge in the printer.

Inkjet papers are paper types with surfaces finished for rapid absorption of the ink that an inject printer bombards the paper with at high frequency in tiny droplets in an inkjet printer. They help prevent the ink from running.

Poster papers are papers for large-format posters, usually wood-containing and heavily sized. These papers are un-coated papers with special properties that allow them to

---

[2]   So called because they were first used to print Bibles around 100 years ago.

soften before being posted and are particularly weatherproof
and water-resistant. Water-resistant papers are highly tear-
resistant even when wet. This property is achieved by special
additives to the fibre stock mixture for the paper. If the additives are also alkali-resistant they are also lye-proof.

LWC papers are high quality, thin, coated papers used
mainly in web offset printing and in gravure printing for
magazines and catalogues.

Electronic paper

The desire to combine screen reproduction with the superior
legibility of printed paper has led the Massachusetts Institute
of Technology (MIT) and various producers to develop so-
called electronic papers.

For example, one variant of **electronic paper** has tiny
spheres with different colours on different sides. These turn
with the aid of electrical fields, and thus make it possible to
see different colours.

Another electronic paper contains small transparent
capsules filled with dye and white particles. When an electric
field is created, these particles float upwards and make the
surface of the paper look white. At other times the effect of
the dye predominates.

Checklist: Choosing paper

- What effect is intended?
- How is the paper to be used?
- How long does the paper have to last?
- How much can the paper cost?
- What image or text quality is the paper to reproduce?
  Note: The maximum quality of reproduction that can be achieved is reduced in accordance with the paper type (see below).
- What printing process is to be used?
  Note: Special papers are needed for most digital printing processes.
- How is the printed product to be handled subsequently?

Colour reproduction on paper types according to ISO 12647

The international standard ISO 12647 classifies the various paper types according to their colour reproduction in printing. If colour management is to be used already in the design and repro process, it is recommended to simulate the ultimate printed product on the monitor and on a proofing system.

The main paper types under ISO 12647 are:

- Glossy coated papers (paper class 1)
- Matt coated papers (paper class 2)
- LWC papers (paper class 3)
- Uncoated papers (paper class 4)
- Uncoated yellowish papers (paper class 5)
- Newsprint.

Reproduction

Reproduction, also called **pre-press** or **pre-print**, is a component of the printing process; it covers all of the working stages required for the original material to be printed to become the **printing copy**. These stages include, for example, page layout, composition and picture processing as well as sheet assembly and the making of printing plates, where applicable.

Correction marks

Certain signs and rules have established themselves in the printing world for correcting texts. Correction marks are laid down in the **international standard ISO 5776** ("proof correction marks"). The German standardisation system **DIN 16 511** is close to this international standard while the British standard **BS-5261 2** deviates somewhat more markedly from it. The current standards have been in revision for some time, as they can be used only for correction on paper, but not for short notes in the electronic workflows that are the current practice, e.g. for comments in PDF files.

Tip: *Corrections must be made clearly. Every mark in the text must be repeated in the margin. The change required is to be written next to the repeated correction mark.*

The most important correction marks according to the international standard ISO 5776 are listed in the following summary.

Marking in text	Mark	Instruction
Superfluous character	/ ₰	Delete character
Word will be finally deleted Word be inserted	⊢ ₰ ⌐ will	Delete word Insert word
Missing space	⌐	Insert space
Del ete space	⌢	Delete space
If the space is too wide	⌃	Space too wide
If the space is too narrow	Y	Space too narrow
An _emphasis_	— italic	An emphasis
Last line of paragraph. First line of new paragraph.	⌐⌐	New paragraph, break in sentence
Last line of paragraph. First line of new paragraph.	⌒	Add to previous paragraph
At the end of a line and at the beginn- ing of a new line, word division changes.	/ ₰ 7 ni	Change word division
Correct lettar	/ l	Correct letter
Change the sequence words of	⊓⊔	Change sequence of words
Change the sequence of lettres.	‖ er	Change sequence of letters
A paragraph ends here. Indent the first line of the new paragraph.	⊏	Indent
An incorrect indent ⊢must be corrected here.	⊢	Delete indent

Colour management

Colour management serves various purposes. It is used by agencies, repro firms and printers to ensure that colour reproduction is consistent throughout the production process – from input, via display on monitors, to output.

Colour management is also used when several people are working together to ensure the unambiguous identification of colour when picture data, documents and PDF files are exchanged.

Colour profiles are very important here. Each colour profile translates the CMYK and RGB colour values into the neutral **LAB colour model**. In order to switch colours from a source to a final colour model, two profiles are linked together via the LAB colour model. The LAB colour model serves as an interface between input, working and output colour models.

Colour profiles fall into the categories described below.

Input profiles define the colour models used by scanners or digital cameras. **RGB working colour models** are used here to exchange and process RGB images. Almost all of today's digital cameras deliver data that has already been colour optimised in a normal RGB colour model.

The most widely used RGB colour models are **sRGB** (fig. 11) and **AdobeRGB** (fig. 12). The sRGB standard corresponds with the colour model of the average monitor. This is the ideal RGB colour model for internet and office use. However, sRGB does not cover some cyan and green colour areas that can be represented in offset printing on coated paper. For this reason, AdobeRGB is better suited to print production, as it can display more colours than sRGB. RGB working colour models are ideal for RGB colour adjustments in programs for image processing, graphics and layout.

CMYK working colour models for printing based on **ISO 12647** have various functions. They are used to simulate the ultimate printed product on a particular paper type on the monitor and in a proofing system. They also control the optimal transformation from RGB to CMYK for subsequent printing and support colour identification in CMYK images, editable documents and PDF files when they pass between several people or companies.

CMYK working colour models for printing based on **ISO 12647** are ideal for CMYK colour adjustments in programs for image processing, graphics, layout, creating PDFs and for making proofs (figs. 13, 14).

Monitor profiles (fig. 15) define the colour model for the monitor.

Printer profiles (fig. 16) define the colour model for a printing system for the paper type and print settings used.

Profiles for printing based on ISO 12647-2. Traditionally, printing has been different all over the world, though the international standard ISO 12647 means that considerable assimilation is taking place. The use of standard profiles for printing on different paper types also differs from country to country.

As the International Organization for Standardization (ISO) does not issue its own profiles for ISO standards, there are various firms and organisations working in this field. The best known include Adobe with standard profiles in its Creative Suite, the European Color Initiative (ECI), SWOP and GRACoL in the USA and IFRA for newspaper printing world-wide.

Many suppliers of proofing systems offer preset installations with profiles from ECI, SWOP und IFRA. The following summary shows the most common profiles for printing based on ISO 12647-2 in Europe and the USA.

Paper types following ISO 12647-2	European profile	USA profile
Coated	ISOcoated_v2 (ECI) Replaces ISOcoated (ECI) and Europe_ ISOcoated_FOGRA27 as of April 2007	GRACoLcoated
LWC (especially for web offset)	ISOwebcoated (ECI)	SWOP
Uncoated	ISOuncoated (ECI)	–
Newsprint	ISOnewspaper26 (IFRA)	ISOnewspaper30 (IFRA)

Establishing a continuous colour management workflow

The essential requirement for a continuous colour management workflow is that the same **colour settings** should always be used for image processing, graphics and layout programs, as well as for PDF and proof generation programs. This applies particularly to CMYK colour settings, which must be appropriate for the paper ultimately to be used in printing.

An individual profile should be set up for each monitor with the help of a measuring device; the proofing system should also be calibrated regularly using an instrument of this kind. Fig. 17 shows colour profiles linked for CMYK separation, for monitor display and for printing.

17

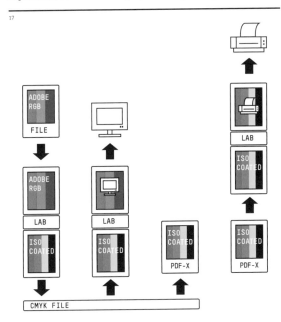

Tip: Use the ISO standard settings in programs. The colour models ISOcoated for Europe and SWOP for the USA are already installed by the manufacturer in the Adobe Creative Suite 2 and in many proofing programs. It is thus possible to set up a continuous colour management workflow with very little effort.

Tip: Colour profiles for ISO printing standards should be used in communications. To achieve comprehensive colour management when different people or companies are working jointly on a printing project, they should agree on the colour profiles for the type of paper to be used. If the work is to be produced in Europe on coated paper, all the RGB data should be separated using the ISOcoated profile and the ISOcoated colour model should be simulated in proofs. The printer should be notified beforehand that the printing will be done according to the ISO standard.

<u>Potential problems arising from colour management</u>

Even though colour management using colour profiles makes things considerably easier in many cases, incorrect colour settings in applications can trigger unintended colour transformations in images, graphics or textual elements. This could then lead to incorrect colour display on monitors or printers.

Tip: Any designer who is not sure that he or she can handle colour profiles in the applications being used for a particular project should use a repro firm for separating RGB images, and under certain circumstances for final artwork, PDF generation and proof.

Halftone screens

In reproduction, a **screen** is an area containing small geometrical shapes, arranged regularly or randomly. Although these shapes can also be round, typical types of halftone screens are those with elliptical dots (fig. 18), lines (fig. 19) (→ p. 174, effect screen) and those with square dots (fig. 20).

Halftone screens are used in printing technology for translating **halftone images** into the black-and-white or full-colour form required for printing by varying either the size or the frequency of the elements according to the image brightness.

A **dot** is the smallest **halftone element** in an image. Several dots together produce a pixel. A colour depth of "8 bit", for example, means that three layers with 256 dots each are superimposed. Unlike a vector graphic (→ ch. Digital Media, p. 110), a **halftone graphic** (fig. 21) is made up of many dots arranged on a fixed grid.

Screen types

The **screen frequency** is the measurement for the number of halftone cells per unit of length. The usual specifications are **L/cm** (lines per centimetre) and **lpi** (lines per inch). The following table shows the approximate conversion values from L/cm to lpi (→ p. 189, screen frequency and screen angle meter).

Conversion of halftone frequency from L/cm to lpi										
L/cm	20	25	30	34	40	48	54	60	70	80	100	120
lpi	50	65	75	85	100	120	133	150	175	200	250	300

The lower the halftone frequency, the larger the halftone cells (fig. 22: 20 L/cm, fig. 23: 60 L/cm, fig. 24: 80 L/cm) and the coarser (less) the details of the graphic image elements will be.

Coarse screen	up to approx. 32 L/cm
Medium screen	from 40 to 70 L/cm
Fine screen	60 L/cm, 70 L/cm and 80 L/cm
Finest screen	over 80 L/cm

Conventional screening is amplitude-modulated screening. The **amplitude modulated screen (AM screen)** – also **autotype** or **conventional screen** – is a screening process in which the halftone dots are the same distance apart and the halftones are created by using **halftone dots** of different sizes (fig. 25).

26

When using this screening process, the colour forms have
to be shifted in relation to each other at fixed screen angles
(→ p. 175) when they are being printed together to avoid
creating moiré effects (fig. 26).

*Tip: Advantages of the conventional screen: it is the most
widely used, and proves best for reproducing the qualities of
previously printed material. It is possible to make a wider
range of colour corrections on the press, as the halftone dot
admits larger colour fluctuations (greater printed area); grey
areas look more even – provided that printing takes place on
a screen with at least 70 lines. Disadvantages: larger colour
fluctuations can occur. It is harder to control colour evenly in
a production run; higher dot gain, more ink used; halftone dots
are visible in the printed form (exception: in the case of very
high halftone frequencies like for example the 90-line screen);
moiré effects (especially in skin tones), saw-tooth effects in
halftones.*

Effect screens are special screen forms and structures such
as grain screens, line screens, circle screens or cross-line
screens.

    The **frequency modulated screen (FM screen)** is a
high resolution effect screen (fig. 27). Frequency modulated
screening is a process for halftone simulation using halftone
dots of the same size. The number of dots in a particular
field determines the colour tone. The FM screen is of very
high quality, but requires greater care in printing and a different working approach for printing plate copying and in the
printing process.

*Tip: Advantages of the FM screen: extremely detailed printed
product through the "simulation" of photographic grain; minor
colour fluctuations; no visible half-tone dots so no undesirable
moiré effects (interference effects); separated colours look like
special colours. Disadvantages: the details "expose" poor or
imprecise image processing; coloured areas can look patchy.*

*Tip: The FM screen is well suited for producing high quality on
coated and cast-coated papers.*

Screen angle

In conventional screens, the screen angle indicates the screening direction, which is measured from the vertical. The human brain finds it easy to perceive angles around 0 and 90 degrees; diagonal placings of 45 or 135 degrees are customary for single-colour reproduction. In multi-colour printing, different screen angles are chosen for the various colours. This avoids superimposition effects (moiré) (→ p. 174, fig. 26). In offset printing, the angles 0, 15, 75 and 45 degrees are used for the four colours yellow, magenta, cyan and black (figs. 28, 29) (→ p. 189, screen frequency and screen angle meter).

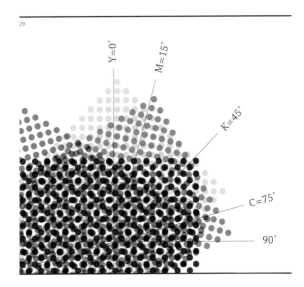

Artwork and control

Artwork is the preparation of the layout (→ ch. Design, p. 41) to produce a definitive ready-to-print copy. The artwork involves examining the material produced so far and optimising it where necessary. This step is closely linked with the subsequent processes and also takes account of features like the type of printing, the binding, and other finishing processes for the final product. Good artwork can be produced only in dialogue with the printer and/or the bookbinder. As a rule, the final artwork product is a file, e.g. an InDesign file or a PDF file (→ p. 179). It is compiled at the end of the graphic design process.

    The artwork files format the final version of all the elements to be printed (e.g. type, colour areas, images). It also sets various technical printing parameters such as colour mode and trapping.

Trapping

Flashes are unprinted white areas between adjacent areas of colour in printed products where the paper is visible (fig. 30). This phenomenon occurs because the press is incorrectly adjusted or the shape of the paper has changed during printing. The risk of flash formation can be reduced by **overprinting** areas or slightly **overlapping** them. This process is known as trapping.

    Trapping should always be used when two different colours (other than white) are directly adjacent to each other. Overlapping is achieved by slightly enlarging lines or areas in relation to the colour areas next to them (fig. 31). The lighter colour is usually overlapped.

    Parts of an object that are **hidden** under other objects can be eliminated. This prevents one colour mixing with another when this is not required. For example, if an area of yellow is printed over an area of cyan, the two colours will mix to produce the colour green (fig. 32). If the top area is to remain yellow and the background cyan, then the yellow area in cyan must be eliminated (fig. 33).

Overprinting is the opposite of cut-out. The only colour that should be set to "overprint" is black. If overprinting is used, the colours underneath are not eliminated, but printed over.

Tip: *Perfect, stitched or stapled binding have different* **bleed** *requirements. For perfect binding, it is extremely important that each page is treated individually – the pages are trimmed at all four edges. The inner edge is milled (→ p. 201, block gluing) and glued. Consequently, pictures that go into the gutter are also bled there so that flashes can be avoided.*

<u>Adding control devices</u>

Feeding marks are printed marks on the edge of the sheet. They show on which page the printed sheet it to be placed so that it can be folded and trimmed correctly for its position.

 Register marks – also called register crosses (fig. 34) – make it possible to check the register accuracy of the individual colours when they are printed together. Register marks are positioned in exactly the same place in every single colour separation (CMYK or spot colours). If the printing is accurate, the marks in the printed work will be precisely on top of each other.

 Trimming marks (fig. 35) indicate how the sheet should be cut to the right format.
 The printing position should always be marked. This is essential for all products that have to be further processed untrimmed or with separation cuts, and for which the type area (→ p. 187, keeping to register) of the printed product has to fit on top of another one. The printing position is established with a set of marks that provide precise orientation for placing the sheets with complete register accuracy in sheet printing (→ p. 178, fig. 37).

 Sheet signatures (fig. 36) are also essential when working on book blocks for establishing the right sequence of sheet parts and staple positions.

37

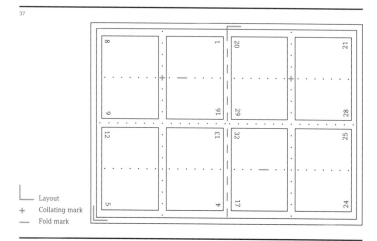

└─ Layout
+ Collating mark
─ Fold mark

Colour reproduction

Separation – also called **colour separation** – converts the colour information from an image into colour data. A colour separation is the proportion of colour in digital copy that corresponds with one colour in colour printing. So in four-colour printing four colour separations are required to make corresponding printing plates in cyan, magenta, yellow and black (figs. 38–41). The individual colour plates are known collectively as a **colour set**, i.e. the sum of all the individual colour separations that come together to produce an end product in colour (fig. 42). **Colour reproductions** can be built up according to a number of principles – achromatic or chromatic.

UCR (Undercolor Removal) is used when an image is being colour-separated, and prevents too many colours being printed on top of each other. When producing colour sets for four-colour printing, identical proportions of the three chromatic colours cyan, magenta and yellow that join in an ideal case to produce a greyscale value are replaced by appropriate proportions of the colour black. This considerably reduces ink consumption. In this way expense is spared, and the printing process, particularly the grey balance, becomes more easily manageable.

GCR (Gray Component Replacement) goes a little further. Here the colour ratio is reduced in a number of colours, so GCR has a greater black ration then UCR. The transitions are fluid and not precisely defined.

ISO standard colour reproductions. If RGB data are converted with a colour profile for ISO printing standards (for example for ISOcoated, SWOP, ISOuncoated, ISOnewspaper), this will give the best possible separation for a particular kind of paper when printing. Ideally, the same colour profiles should be used for this separation and the subsequent proof.

Imposition

Printed products consisting of a large number of pages are not printed as individual pages; instead, several pages are placed on the printed sheet together. As a rule, a printed sheet contains 4, 6, 8, 12, 16, 20, 24, or 32 pages that are placed together on a plate, or form (→ p. 183). The pages are arranged according to a particular scheme.

Imposition means positioning matter to be printed on a form with the pages in the correct position.

The so-called **imposition scheme** fixes the page placement and the position of the folds. Page numbers vary according to the total number of pages and their arrangement (fig. 37) on the individual **imposition forms**.

Reproduction control

The final artwork is usually created as a **PDF** (Portable Document Format); printers use the PDF directly as copy for making the plates.

"**PDF/X**", an ISO standard, has now been established for digital printing copy. The PDF format offers distinct advantages at the pre-press stage. It is easier to carry out jobs such as pre-flight, trapping (→ pp. 176, 177) and imposition at this stage, and to check them in a PDF than in other data formats.

The so-called **pre-flight check** is made before **exposure** (film or printing plate), to simulate the **output process**. The files are checked in terms of various parameters such as format, resolution, fonts (→ ch. Typography, pp. 90–92), colour mode, separation, feeding marks (→ p. 177) or trapping (→ pp. 176, 177).

Tip: *A number of pre-flight programs are available on the market at the time of writing: e.g. "PitStop" or "FlightCheck" by Markzware and the integrated pre-flight function in Adobe Acrobat.*

Proof and machine proof

The **proof** is an important instrument for checking quality in the printing process. Its greatest advantage lies in the fact that, independently of the preceding pre-print process, it can provide a reasonably realistic impression of the finished printed product.

A proof is made to do a binding check on the colours in a printed product in advance. It is less demanding in terms of time and cost than a machine proof. It is essential to control the pre-press phase (→ p. 167) with a colour management system (→ pp. 169–170) if an effective proof is to be made.

The so-called **machine proof** comes closest to the subsequent printed product. This is made on a press from the printing plates and with the paper to be used for the actual print run of the job.

An even more precise result will be achieved by using not just any **press** but the one that is to be used for the print run later. Here it is also important to use the same paper and the same inks or varnishes as will be used for printing. A print made in this way will give a clear idea of the quality of the finished product as the production conditions are identical.

A so-called **colour scale** appears on the proof, in order to assess the colours on the printing material. In four-colour printing (→ p. 186, CMYK) it consists of the four process colours cyan, magenta, yellow and black printed alone and in various combinations on a small area (→ p. 182, fig. 43).

The **blueprint** is a monochrome print of the completed print copy. Today, the term blueprint can also refer to an equivalent simple black-and-white printout.

In the **analogue proofing process** (e.g. Cromalin, Matchprint) the proof is made using film that has already been exposed. The analogue proof largely matches the subsequent printed product. Analogue proofs are hardly used today, as films are increasingly becoming less important as a basis for plate exposure.

The **layout proof** is a printout from computer data that is not binding in terms of colour. It is used mainly for checking mechanical details (e.g. content, layout, position). Printing systems for layout proofs are usually laser colour prints or simply inkjet prints. In many cases it is impossible for a printer to reproduce the colours from a layout proof in offset printing. If colour is to be legally binding between client, agency, pre-press company and printer, then an accurate proof – for example a digital proof – is essential.

The digital proof

Producing a **digital proof** requires a proofing system consisting of a high-quality inkjet printer and proofing software. A proofing system is regularly calibrated with a colour measuring device and the proofing software. It is also important that the people producing the printing data and the printers themselves agree about a colour standard at an early stage, so that this can be simulated in the proof. It is easiest to agree on a colour profile based on ISO standards such as ISOcoated in Europe or SWOP or GRACoL in the US.

Control wedge and control line in digital proofs

As digital proofs often circulate between different people and companies, the essential proofing data must be included on the proofs themselves. The control wedge and control line are used to this end. The control line shows which file was proofed, when the proof was made and what printing standards were being simulated. The control wedge makes it possible to use a colour measuring device to check even at a later stage whether the colour reproduction on the proof really corresponds to the simulated printing standard.

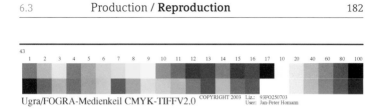

Ugra/FOGRA-Medienkeil CMYK-TIFFV2.0   COPYRIGHT 2003   Liz.: 93FO250703   User: Jan-Peter Homann

<u>Different control wedges world-wide for digital proofing</u>

Different proof system producers and regional organisations usually prefer different control wedges for the digital proof. For international projects it is helpful to agree on a single control wedge. At present, the Fogra CMYK media wedge is the most common world-wide, and is already bundled with many proofing systems (fig. 43).

Tip: *For German users: Digital contract proofs in Germany are obligated to simulate an ISO printing standard and contain a control line and the Fogra media wedge. Further details are laid down in the "Media Standard Print", which can be found in English at www.bvdm-online.de/Aktuelles/Downloads.php.*

So-called colour **plots** are very frequently used today. The printer uses them as a device for checking the completeness, positioning and content of each graphic element. Colour plots are not contractually binding in terms of colour, however (→ p. 181, layout proof).

    Unlike the **hard proof**, the **soft proof** simulates the printed product on a monitor. A conclusive soft proof can only be created with correct colour calibration and mastery of electronic image processing (→ pp. 169–172, Colour management).

<u>Checklist: Artwork</u>

Basic:
- Colour model (colour profile corresponding to the ISO standard in printing)
- Format
- Number of pages
- Bleed

Extended:
- Image data
- Image resolution
- Data formats
- Retouching and colour corrections
- Texts
- Languages
- Make-up
- Trapping
- Colour separation and configuration
- Special colours and separation

Print forms

The **woodcut** is one of the oldest processes for making print forms. It uses long log boards, from which parts are removed on the basis of a picture drawn beforehand with various cutting tools. The cut-out areas are left uninked when printing. The raised parts that remain form the image that will be printed. The earliest woodcuts for duplication on paper can be shown to have existed in China in the 6th century AD.

Today, pre-press printing data for making printing plates are prepared with a **RIP (Raster Image Processor)**. Its most important function is to create the screen (→ pp. 172 – 175) for printing images and other graphic elements. As a rule, a RIP is a computer in its own right, but it also exists as complex software.

Blocks is the old term for the plates (usually metal plates) used for reproducing images in the letterpress printing process (→ pp. 191, 192). For halftone material (→ pp. 172 – 175), **halftone blocks** used to be prepared, and for line material, **line blocks**. Today the term "block" is mostly used in flexo printing (→ p. 192) to describe letterpress forms or also in pad printing (→ p. 194) to describe the gravure forms. In addition, the stamp used for hot foil stamping is often called a block.

Two basic processes are used to make forms or plates for offset printing: computer-to-film and computer-to-plate.

Computer-to-film (CtF) is a process for making films for plate manufacture when imposition is to be carried out electronically, but the plates are still made photo-mechanically. In the computer-to-film process data from various sources is consolidated electronically and then put onto film. Fig. 44 compares the computer-to-plate, computer-to-film and computer-to-print processes.

A more recent variant for line or halftone material is **desktop computer-to-film**. Here, the film used for making the plates is not produced photographically, but printed. It is essential that the printer (laser or inkjet) can print the film accurately to size.

In **computer-to-plate** (CtP), data are transferred directly onto the plate from the computer, without film as a transmission medium. Most printers currently use this process.

Computer-to-print is a process that does not use plates at all. The machines use methods similar to laser printing to print directly from the data prepared for each commission. The process is often also known as digital printing (→ p. 195).

Tip: *Computer-to-print is suitable for small print runs and easy to use for personalised printed matter.*

Direct Imaging is a technology in which postscript data from the pre-press stage are used to produce all the print forms at the same time with complete register accuracy. Here, screen data delivered by a RIP (→ p. 183) drive infrared laser diodes. These create tiny indentations in a print film with an ink-resistant surface, thus revealing a layer that carries an ink-receptive layer. The result is a printing film that can be used for waterless offset printing (→ p. 190).

Cross media is the name used in printing for using the same printing data several times for a wide variety of media. This means that the same data can be used to produce printed pages, material on CD-ROMs and internet pages. The PDF (→ p. 179) is an important format for cross media because it can be used to store print-quality documents and it can interact with other formats and programs.

44

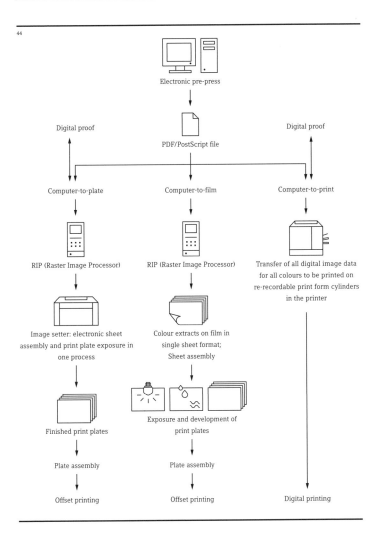

Electronic pre-press

Digital proof

PDF/PostScript file

Digital proof

Computer-to-plate

Computer-to-film

Computer-to-print

RIP (Raster Image Processor)

RIP (Raster Image Processor)

Transfer of all digital image data
for all colours to be printed on
re-recordable print form cylinders
in the printer

Image setter: electronic sheet
assembly and print plate exposure in
one process

Colour extracts on film in
single sheet format;
Sheet assembly

Finished print plates

Exposure and development of
print plates

Plate assembly

Plate assembly

Offset printing

Offset printing

Digital printing

Printing

<u>Printing inks and properties</u>

The **standard colours** in four-colour printing are cyan, magenta, yellow, black, also **CYMK**, and are called **process colours**. Here CYMK stands for cyan, magenta, yellow and key.

In **seven-colour printing**, violet-blue, green and orange-red are added to the process colours. This system was developed according to a theory by Harald Küppers (→ ch. Design, pp. 10, 11), and delivers a luminous printed product that could not be achieved with four-colour printing.

Printing inks to **ISO 2846** are ideal for achieving the best possible printed result using ISO 12647. ISO 2846 has replaced the earlier European scale in Europe, and is now a global standard for printing inks.

Special colours – also called **spot colours** – are usually used for colour tones or colour effects that cannot be realised with process colours. Examples of these are fluorescent colours, gold and silver, as well as some colour shades in the Pantone series.

Pantone colours are based on a globally used system of standardised colours introduced for the print media industry by Pantone, Inc. in 1963. The Pantone System provides reference colours mixed from fourteen basic colours, black and white. There are over 1,100 Pantone colours available today.

The **HKS** system is a hybrid system used mainly in Germany. It is based on nine basic colours plus black and white.

RAL colours are standard colours according to a series of colour collections for the industry, published by the German Institute for Quality Assurance and Certification.

The colour values from these various models can be converted into CYMK. Because the colour profiles differ, only approximate values can be achieved.

Printing inks must have certain **properties** and meet certain requirements if they are to do justice to a broad range of uses. In general, they must not rub off or smear, and be colour stable. Colour stability defines an ink's ability not to react to the **influence of light** from the natural solar spectrum.

So-called **coldset** inks are used for newspaper printing in the web offset process. These inks dry exclusively by **soaking** into the paper. **Heatset** inks are inks that are essentially dried by being heated briefly at the end of the printing process. This is done with hot air at temperatures of 120 to 150 degrees Celsius. Heatset colours are used in web offset printing.

Structural or **physical inks** create shimmering colour effects that vary according to the angle from which they are looked at. The colour impression is not created by the pigments alone, but also by their physical structure. These inks contain special structures such as thin, transparent platelets, for example, that selectively reflect light on certain wavelengths using interference effects.

UV inks harden when irradiated with ultraviolet (UV) light. In addition, these inks do not contain any volatile substances, but contain colour pigments and also individual molecules and short molecule chains that can combine to form polymers, together with so-called photo-initiating agents. The latter disintegrate when irradiated with UV light, thus forming highly reactive fragments. These radicals trigger a polymerisation process that creates solid, three-dimensional net structures (→ pp. 197, 198, varnishes).

In UV offset printing in particular (→ p. 190), UV inks make it possible to print on far more materials than just paper. With UV inks it is also possible to print a true opaque white on coloured print media, or to print with gold- or silver-like inks.

Printing process control

Control aids form the basis for effective quality control in pre-press and printing. They form part of a comprehensive quality assurance system. During printing, the print sheets are checked with a **linen tester**, and now electronically as well, by **optical sampling**. So-called **misregistrations** lead to lack of sharpness in reproducing images and print, flashes (→ p. 176) and colour deviations under some circumstances.

Keeping to register. For first run and back-up printing, the printed pages have to be printed with accurate registration. The **front** of the printed sheet is printed on the first run, and the **back** on the back-up run. If the first run and the back-up do not match, there will be fluctuations in the type area.

The **ink absorption** (the quantity of ink absorbed by the paper during printing) can have an effect on print quality and has to be adjusted during the print preparation period.

Colour density (the amount of colour per area unit) is the optical density of areas printed in colour. This value can be measured with special equipment like a **densitometer**, for example. Here, the colour densities of individual shades are compared with each other. Densitometry measures tonal values, but not colour shades. Densitometry also plays an important part in photography.

The **colour wedge** is a measuring strip used for printing density. It consists of full tone areas for each ink colour (→ p. 186). **Grey wedges** display the colour ratios needed to achieve a neutral grey – the **grey balance**. The grey balance is achieved with equal ratios of cyan, magenta and yellow.

The **characteristic printing curve** is a diagram representing the dot gain of a press for various surface coverages. It is a graphical representation of the link between the **tonal values** of the pre-press products, e.g. the tonal values of screen data, film or printing plates, and the equivalent tonal values in printing. The so-called **dot gain** is derived from

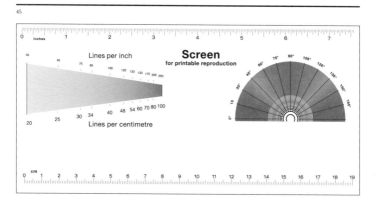

45

this. The characteristic printing curve identifies how much
a screen image darkens as a result of dot gain. A screen
area becomes darker as a result of dot expansion and light
capture. Proofs simulate the dot gain for various paper types.
The necessary parameters for dot gain on different paper
types are laid down in ISO 12467.

     A **screen frequency** and **screen angle meter** (fig. 45) can
be used to measure screen frequencies and screen depths in
pictures that have already been reproduced. A **screen wedge**
is a control field used to determine the characteristic printing curve of a press. It consists of colour fields, usually in ten
per cent stages, representing colour shades from nought to
one hundred per cent.

Printing processes

Today, a distinction can be made in principle between two
basic printing processes: non-impact, or **plateless printing**
and the **impact printing process** with plates. The overwhelming majority of print production is now carried out
on printing plate systems (**impact systems**), the most common of which is offset printing. Impact systems also include
the **flat-bed** (→ pp. 189–191), **letterpress** (→ pp. 191, 192),
gravure (→ pp. 193, 194) and **screen** (→ pp. 194, 195) printing processes. **Non-impact systems** include, for example,
electro-photographic processes such as **digital printing**
(→ p. 195) or **inkjet printing** (→ p. 196).

Flat-bed processes

The inventors of **offset printing** (→ p. 190, fig. 46) were Ira
W. Rubel (USA) and Caspar Hermann (Germany). This process is based on the mutual repulsion of fat and water. Offset
printing ink is fat-based. The areas to be printed on a plate
are both **lipophilic** (attract fat) and **hydrophobic** (repel
water) and thus take in the ink. The areas that are not to be
printed on a plate are both lipophobic (repel fat) and hydrophilic (attract water) so that the ink is repelled.
Printing and non-printing areas are on the same plane. The
non-printing areas are first moistened with a thin film of a
moisturising agent by a **dampening unit**. After that the ink is

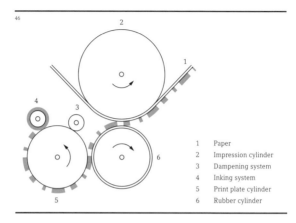

1   Paper
2   Impression cylinder
3   Dampening system
4   Inking system
5   Print plate cylinder
6   Rubber cylinder

transferred to the printing areas by the ink form rollers in the inking unit. The printed image is first transferred to a **rubber blanket cylinder** (via the blanket) and then from there to the printing substrate (**indirect printing process**). Fig. 47 shows a typical characteristic of offset printing, the smooth edge.

Another form of offset printing is **waterless offset printing**. This process prints without water, using specials inks, plates and cylinder coatings (silicon). Because the paper is moistened less, a finer printing screen and more precise ink application are possible. Waterless offset printing is more environmentally friendly than the traditional offset process, as it works without dampening agents.

UV printing is a special form of offset printing. This process uses presses with a built-in varnish station and UV drying. The process makes it possible to achieve high-quality offset printing on polyester, PVC, PET, polycarbonate, metallic foil and vinyl as well as other impermeable surfaces. The forerunner of modern offset printing is **lithography**[3] (flat-bed process), which was invented by **Alois Senefelder** (1771–1834) in 1789. This process is used to make printing forms for printing with stone. The artwork is transferred directly to a smoothly polished block of limestone with special ink or chalk. The stone block is then moistened before it

[3] Ancient Greek: lithos = stone; graphein = writing

is inked with oily ink. The written areas attract the oily ink, while the unchanged limestone repels it. Lithography is now essentially used only for graphic art work.

The term **lithography**, **litho** for short, is now used for copies (positive, laterally inverted **line or screen films**), and also for artwork (e.g. scan material).

<u>Letterpress printing</u>

Letterpress printing is the oldest industrial printing process. The parts that are to be printed are raised. These raised areas are inked, and when printing takes place, some of the ink is left on the print substrate. A distinction is made between three types of letterpress printing.

Firstly, in the so-called platen press, one surface presses against another surface (fig. 48). Secondly, in a cylinder press, a cylinder presses against a surface (fig. 49). Thirdly, in rotary printing, two cylinders roll against each other (fig. 50).

Today, letterpress printing is used for small print runs and special work (dye-cutting, embossing, perforating, numbering etc.), as well as in some cases for newspaper printing.

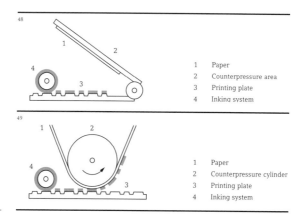

48

1	Paper
2	Counterpressure area
3	Printing plate
4	Inking system

49

1	Paper
2	Counterpressure cylinder
3	Printing plate
4	Inking system

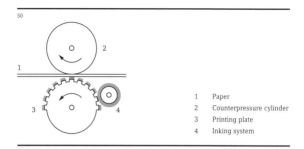

1	Paper
2	Counterpressure cylinder
3	Printing plate
4	Inking system

Flexo printing is another letterpress process (fig. 51: principle of "round against round"). Flexo printing uses **photopolymer letterpress plates** or moulded plates (**rubber plates**). A very wide variety of materials can be printed with low viscosity ink to a screen resolution of 54 L/cm (→ p. 173). Fig. 52 shows a typical feature of flexo printing, the colour rings.

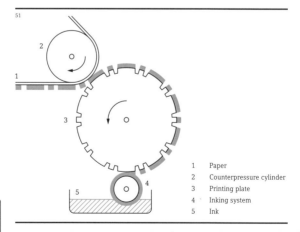

1	Paper
2	Counterpressure cylinder
3	Printing plate
4	Inking system
5	Ink

Tip: *Flexo printing is a rapid printing process that is suitable for the variety of print substrates required when printing packaging.*

Gravure printing

In **gravure printing** (fig. 54), the printing elements take the form of cells (fig. 53). As a rule, the printed image is transferred to the cylinder with a diamond stylus using an **electromagnetic gravure process**. The cylinder is inked completely for the print run. Then a **doctor blade** strips the excess ink off the surface and the ink is left only in the indentations. A rubberised roll presses the paper against the cylinder and the ink remaining in the cells is deposited on the paper (direct printing process). A doctor blade is a thin, straight steel rule in gravure printing. In **screen printing** (→ pp. 194, 195) a doctor blade with a rubber or plastic edge is used to distribute the ink on the screen.

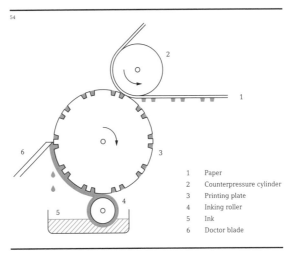

1	Paper
2	Counterpressure cylinder
3	Printing plate
4	Inking roller
5	Ink
6	Doctor blade

Tip: *Because gravure printing is a relatively expensive process, it is only worth using it for print runs of 300,000 prints or more. This printing process can be used to make almost limitless copies.*

In contrast to gravure printing, **recess printing** uses plates in which the print motif is engraved in the form of lines and dots.

Tip: Recess printing is used mainly for prestigious printed products such as securities, postage stamps or banknotes. Here it is possible to apply ink to a thickness of about 0.1 millimetres (the print can be felt as a relief). This kind of raised printing is a means of protection against banknote forgery. Narrow strips of plastic are used, where applicable with breaks that appear as negative printing against the light. The **security strips** *applied to the notes provide further protection.*

Pad printing is an indirect gravure process in which the intermediate medium is a flexible, often hemispherical, silicone rubber pad that transfers ink from the plate on to the area to be printed. This method makes it possible to print on a wide variety of irregularly shaped objects.

Etching developed from **copperplate engraving**. In this process, a metal plate (usually copper) is coated with an acid-proof varnish. A so-called etching needle is then used to expose the ground areas that are later to appear in the printed image in the form of lines, hatching etc. The metal plate is then treated with acid, which etches the metal where it has been cleared. When the varnish has been removed, the plate can be used as a gravure form.

Screen printing

In **silk-screen printing** (fig. 55) – also known as serigraphy – the print form is a fine-mesh sieve fabric with a printing stencil. A doctor blade (→ p. 193) presses the ink through the form on to the substrate. Several methods are used to transfer the image on to the screen. For example, the original image can be copied on to a screen with a light-sensitive layer under UV light. The exposed areas of the layer are hardened; the unexposed ones can be washed off with water, leaving the screen permeable to the printing ink in these areas.

Tip: Ink for screen printing is applied five to ten times as thickly as in other printing processes. It is therefore particularly suitable for high-quality advertising printing, signs, posters and print for packaging, and also for CDs and DVDs. In addition, the silk-screen process can be used to print special colours containing more pigments or for special applications containing glue or coloured areas that can be rubbed off (e.g. on lottery tickets).

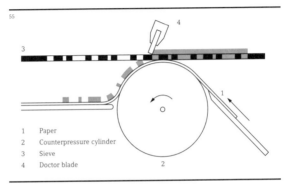

55

1	Paper
2	Counterpressure cylinder
3	Sieve
4	Doctor blade

Other printing processes

Other printing processes have established themselves in recent years. **Electro-photographic processes** are now used in numerous copiers, laser printers and digital printing systems for single- and multi-coloured printing and duplicating.

Digital printing is a process in which information is transferred directly from the computer to paper, without any printing copy (→ p. 167) being produced. Here, colour printing technology is combined with the mechanism of a printing press. Smaller print runs can make financial sense because high offset printing costs, e.g. for plate-making, are avoided.

Tip: *Digital printing cannot achieve the quality of traditional printing processes, but it is quicker and cheaper for small print runs; it also makes it possible to use special techniques such as personalised printing (→ p. 196) or printing on demand (→ p. 196).*

Laser printing is also an electro-photographic process. The substances providing the colour are called toners. The toner is applied to the surface of the paper and fixed there by radiant heat.

Inkjet printing uses **thermal or piezo technology**. The **inks** used for inkjet printing are mainly dyes dissolved in water. They are inclined to run, penetrate or roll off, and when printing in colour the inks can mingle in border areas. The substrate can also distort or become wavy because of the high proportion of liquid in the ink (90–95 per cent).

Tip: *As there are a large number of different printers and inks, it is advisable to make application-specific test prints before actually printing.*

In **personalised printing**, individual copies or parts of a print run are given individual imprints. Personalised printing requires a digital printing process (→ p. 195) for the individual imprints, so that the print data can be changed between the individually printed copies. This process is frequently used to make printed products with addresses or personalised greetings on them (→ ch. Marketing, p. 230, Direct Marketing).

Printing on demand is a scheme with no fixed run for the printed product; small part runs, down to individual copies, are printed as required. Printing on demand relies on digital printing, which makes it possible to print directly from the pre-press data without making forms and setting up presses.

Books on demand are produced and delivered individually when they are ordered.

Tip: *Using printing on demand for small print runs can reduce costs as nothing is spent on storage and unsold copies; it is also possible to produce individual books from prescribed sections.*

Printed products with errors are known as **spoilage**. Broadly speaking, this includes all waste paper in the printing industry, as well as faulty or damaged paper, packaging material, excess printed products and returns.

Post-press

Post-press includes all of the processing steps for printed products that lead to a complete printed product after the actual print run. According to the product type, this includes folding, compiling and trimming the printed sheets as well as binding; post-press also includes special processes such as perforation, lamination (→ p. 198), dye-cutting or punching (→ p. 202). Packing and distribution, if applicable, are also included in this phase.

     If a post-press process takes place directly in the press (e.g. varnishing or folding), or the modules required for this working step are linked with the press, the term **in-line post-press** is also used.

Surface finishing

The varnishing process takes place in the last printing phase. **Varnishing** is the application of a colourless, glossy- or matt-drying layer of varnish. This is done either as **print varnish** in the press, or as water-based emulsion varnish using a special varnishing machine. Varnishing protects the surface and enhances the abrasion resistance of printing inks.

Tip: *Care should be taken with full-surface varnishing as it changes the surface character of the paper, and can change the hue of the printing ink.*

Varnishing can also be used in specific areas as a visual device, as for example in spot varnishing, in which the varnished area is treated like a colour area. **Shadow varnishing** is a half-tone **spot varnishing process**, the results of which are dependent on the density of the printed image. Planed image areas receive more varnish, screened image areas correspondingly less.

     Effect varnishes are special varnishes used to achieve particular effects in printing technology. As a rule, they consist of pigments with particles of varying colours, shapes and sizes. **UV varnishes** are varnishes based on unsaturated polyesters or polyacrylates or a combination of the two; the drying process is triggered by ultra-violet. This lasts only a few seconds, thus UV varnishing is a very swift process.

Tip: UV varnishes produce good results only on coated and cast-coated materials. UV varnishes cannot be bonded and have only limited resistance to scoring and creasing (→ p. 200).

Water-based varnishes – also emulsion varnishes – are usually applied in varnishing machines, and in some cases also in the press' own inking system. The varnish layer can be up to three μm thick. It is not as glossy as UV varnish.

Tip: Water-based varnishes dry quickly as the water evaporates; they are odourless, and yellow very little.

Fragrant varnishes are a specialised product. They contain fragrances in micro-capsules, which are released when the marked areas are rubbed.

    Lamination is also a finishing process. A layer of **transparent foil** (usually in polyester) is applied to the printed product to protect it, or to make it look more attractive. For packaging liquids, the industry uses foils that are also intended to protect the aroma of the product. There are different foils, e.g. glossy, matt or with slight embossing.

    In **hot foil stamping**, a thin film of plastic is applied to the substrate under pressure and heat instead of ink in the letterpress process. Form punches are etched or engraved for this purpose. A layer of aluminium is vapour-deposited onto foil in a vacuum to create a silver gloss. A gold or copper effect can be achieved with a coating of yellow or pink varnish. Hot stamping foils can even contain structures (holographic foils).

Tip: A brass punch is recommended for large print runs as well as for some materials and punching motifs. The brass punch will last longer, transfer the heat better and achieve a particularly bright result; however, it is considerably more expensive than a magnesium punch. A magnesium punch can be etched or milled for short print runs. The punch (sometimes called the block) is usually made of rubber or plastic for cold stamping.

Blind stamping involves stamping paper with a gravure and a matching opposed template under high pressure. Stamping with a raised motif is called embossing, while an image depressed below the surface of the paper is debossed. A motif can acquire various levels, for example, by stamping in several stages. Three-dimensional effects can be created by relief stamping. Embossing can also be simulated by the application and fusing of special powder materials.

Folding

In **bookbinding**, the **fold** is a sharp crease in paper. Folding produces either a product with continuously numbered pages or a smaller format. Folding machines are usually used for this purpose. As a rule, a distinction is made between **cross folding** and **parallel folding**.

In cross folding the next fold is always at right angles, and in parallel folding, parallel with the preceding one. These two basic types of folding can also be combined.

The simplest kind of folding is **single folding**, where the original sheet (→ p. 155, untrimmed formats) is folded only once (fig. 56). Repeating this process also makes **double and triple folding** possible (figs. 57, 58).

The folded printed sheets are assembled and combined. The joined edges of the paper are opened up by trimming. The pages are then only connected by the fold at the gutter. Folding techniques include **altar**, **parallel** and **concertina** folding.

An **altar fold** – also gatefold – is folded so that two wings are created to be opened out (fig. 59).

Zigzag or **concertina folding** is a continuous parallel fold in which each parallel crease is folded alternately in the opposite direction – like a concertina (fig. 60).

Parallel folding is a style in which each fold runs parallel with and in the same direction as the inner part of the sheet. The number of folds determines the end product. A double parallel fold produces three sheets and six pages, and a triple parallel fold four sheets and eight pages.

Scoring and creasing

Scoring is the appearance of indentations on the back of the fold. To prevent high grammage rigid materials from breaking when folded, the folds are **scratched** (fig. 61), **scored** (fig. 62) or **creased** (fig. 63). The scoring or crease should always be applied to the inside of the sheet to be folded (fig. 64: correct, fig. 65: incorrect).

Tip: To avoid cracking at the joint because of strain, it is recommended that the book cover (→ p. 202) is laminated before being covered with paper.

Trimming

Trimming gives a book block, a brochure or a magazine a smooth edge or book trim. The product is trimmed on the three sides on which it is not bound. In perfect binding (→ p. 201) all four sides are trimmed. Trimming also separates the individual pages from each other, which makes it possible to open the book, brochure or magazine.

The **trim** is the paper edge of a printed product that protrudes beyond the dimensions of the finished product. The trim enables all the pages of a book or brochure to be trimmed to the same size.

If, as in the case of stapled products like magazines, a number of sheets are placed inside each other, the inside ones usually protrude more than the outside ones. To achieve a straight edge a so-called **front cut** reduces all overhangs to the same size.

Tip: For front cutting it is important even at the design stage to note that, when a large number of pages are being stapled, the inside pages will be more heavily trimmed than the cover, for example. Flaps and inserts should be about 1–2 mm smaller than the final book block format, as otherwise they will be cut into.

Trimming errors (fig. 66) on bled images or areas lead to flashes (→ p. 176). This is why bled images or areas are usually placed with 3 mm of additional trim.

Binding processes

For **stitched** or **stapled binding**, the printed matter is bound to form a book block or other product using thread or wire staples, respectively.

Stapling – also **wire binding** – is a common type of binding. For back stapling, folded sheets are placed inside each other with a paper or card cover and fastened together from the outside through the back with metal staples (fig. 67).

For **block stapling**, the trimmed sheets are fastened together on one side with staples (fig. 68).

Note: Block stapling can be used only up to a certain thickness.

For **stitched binding**, the oldest of these binding types, the sheets are stitched together through the back, using single or double stitching. In book manufacture, stitching is the highest quality and best-wearing binding type (fig. 69).

Spiral binding (fig. 70), **plastic binding** (fig. 71) and **wire-o binding** are loose-leaf binding systems. Here stacks of paper are punched at one side and the appropriate binding is inserted.

Perfect binding is **threadless binding**. The simplest kind is **block gluing**. The spine of the book block is trimmed or milled to produce individual sheets, which are then bound with adhesive. Various kinds of adhesive are used for this purpose, like for example "Dispersion PUR" or "Hotmelt". In block perfect binding the fixed book block is glued (fig. 72).

Tip: If very smooth paper with extensive areas of print is to be perfect bound at the spine, then working with PUR is recommended. The so-called pull test can be used to test the binding: a sheet in the block is pulled. If the page tears rather than being pulled out, the perfect binding has passed the test.

For **fan binding**, the protruding gutter is fanned out to one side and covered with adhesive after the block has been clamped (fig. 73). The process is then repeated by fanning out the other side of the block. This gives the adhesive more "grip area" and provides more hold for the individual pages. The process is used for single and special products and cannot be done by machine.

Paperback binding

The term **paperback** is used for a binding process in which the cover is glued, stapled or stitched directly to the back of the block with one or more sets of pages. Usually, the cover takes the same format as the book block and is made of card or of the same material as the pages.

A distinction is made between various kinds of paper-back binding, e.g. **Swiss** (fig. 74), **English** (fig. 75), **French** (fig. 76) or **stiff** (fig. 77).

Bound books with their book block fastened to the **cover** with gauze and endpapers are called **hardback** or **hard-cover** books. A book cover (fig. 81) is made up of a front cover, a back insert and a back cover made of cardboard (→ pp. 162, 163) onto which a covering material such as fab-ric, paper, leather, artificial leather or plastic is applied and glued. The cover and the book block are fastened together by the endpapers.

The book block can also be finished with a **decorative band** (fig. 78) and also have a **ribbon bookmark** (fig. 79). A **coloured edge** (fig. 80) is another possible fine finish.

Tip: *An early decision should be taken about whether the book block is to have a straight or a curved back. Note that the first and last pages will always be 5–7mm narrower – according to how far the bookbinder over-glues. This should also be borne in mind for the layout of these pages, so that elements that are placed centrally will always remain centred.*

Selective binding is an individual manufacturing process for bound printed products. For example, catalogues can be produced in several versions from a selection of different components and their content matched to the wishes of dif-ferent customer groups. Selective binding makes it possible to provide individual advertising motifs for various sectional editions of magazines (→ ch. Marketing, p. 221).

Prestigious finishing is the general term for all value-adding post-press processes. As well as varnishing and laminat-ing (→ p. 198), finishing also includes **dye-cutting** (fig. 82), **punching** (fig. 83) and **perforating** (fig. 84).

Production in Commercial Practice

Sequence of work at the printers

1. The PDF data are checked in terms of basic parameters (e.g. format) and technical suitability.
2. If a proof is provided, the printers will check whether this meets contractual proof requirements, e.g. the presence of a control wedge (→ pp. 181, 182) and control strip.
3. The PDF data are converted for the printing process in the RIP (→ p. 183).
4. The processed data are imposed electronically (film presentation is now rare), and the digital sheet data produced are later used to create plates.
5. Plots are made from the outputted print sheets, usually in the original format, imitating the subsequent printed sheet.
6. The plots are assembled and folded in the same manner as the printed sheets will be later. The printers can check again whether the output scheme is correct.
7. The plots are presented to the client for final inspection and approval for press. As the printouts are made from the files that have already been processed in the RIP, trapped objects and type are already in the form in which they should appear later in the printed product. Only the reproduction of colour and images should be viewed with reservations: it is not colour-binding for printing, and the picture data are often printed at a low resolution to save time.
8. The client approves the material for press on the basis of the plots, or corrections are made by the client or authors.
9. The plates are made after the plot has been approved. **Note:** Every correction that is made after this will considerably increase costs, as new plates have to be made after every correction.

<u>Press check and print approval</u>

Print approval is the approval of a print sheet by the client that initiates the **print run**. This print check includes defining **colour densities** for the **production run**, as well as **density settings** and **register maintenance**.

Print approval offers the client a final opportunity to check that everything is in order. However, most corrections at this stage involve additional expense, as new plates would have to be made and expensive machine down-times are involved.

The final print approval is particularly important for **colour matching**, as it is not usually possible to achieve precisely the desired colour for every detail, and the printer often cannot decide which compromises would be acceptable to the customer.

Approval for printing is given as an OK to print, formerly called **imprimatur**.[4] The print run takes place after **approval** (→ ch. Law, p. 246, liability risks).

Tip: *With the control tools available today such as proof and colour plot it is possible for every printing firm to produce a printed result that meets the customer's requirements. Nevertheless, it is often worth being present at the printers for the proof acceptance.*

We have summed up the usual procedure here, taking offset printing as an example (→ pp. 189–191):

The printer starts off with the basic machine set-up, using the colour values that **Fogra** recommends for the paper. Once this is done, a print sheet is presented to the client. This print sheet must be presented under standard D50 light, as the impression made by the colour can vary according to the ambient light (→ pp. 169–172, colour management). The client checks mainly colour. Register correction is a matter for the printer.

For larger printing commissions, the reference colour densities are established using the first printed sheet; these are then used for all subsequent printed sheets. This ensures that the colour impression remains the same throughout a product, for example all the way through a book.

---

[4] Latin: Let it be printed.

These colours should only be deviated from in the subsequent printing process in exceptional cases.

There are technical limitations for **colour correction** in printing. Colour corrections can only be undertaken inside the colour zones that run vertically. As several pages are always printed on a single sheet (→ p. 179, Imposition), each colour correction always affects the pages before and after it on the plate.

As soon as satisfactory print settings have been found, **production printing** starts. Production run is the expression for the print run on the basis approved by the client.

The printer must ensure that the individual colour densities and the colour impression remain constant. The printing process is an unstable one, in which all the elements like paper, ink and water have to be kept in balance. This is usually done automatically today. But the printer constantly examines control sheets during the print run, and corrects register or colour densities where necessary.

Common printing problems

Sometimes technical problems arise in the production run or at the press check stage. As a rule the printer recognizes these in time and can act to stop them at their source. The most frequently occurring problems, the ways of identifying them and possible solutions are listed here.

Hickeys are foreign bodies on the printing plate or the blanket. They interrupt ink distribution. Hickeys cause light or dark patches. The foreign body has to be removed from the plate or the blanket; sometimes the press has to be stopped in order to do this.

Defective blanket. The printed motif is transferred from the plate to the blanket and from there on to the paper. Blankets have to be able to withstand heavy mechanical loads as a result of washing and the printing process itself. This can damage the surface of the blanket, which means that the printed image would not be printed perfectly at these points. This effect is usually identified by a lighter area in the printed matter, and is not easy to spot. Defective blankets are replaced.

Mechanical flaws on the plate often consist of scratches, usually recognisable as thin coloured lines. They are created by careless handling or faults in the emulsion coating on the plate. Often the plate will have to be remade, though sometimes printers can make corrections with a correction pen.

Smears are caused by uneven wetting and ink distribution in the press. The result is that the printed image can no longer be transferred cleanly, as there is too much ink on the plate or the blanket. This blurs the printed image, and shadows appear on the outlines. This fault can be corrected by regulating ink flow and print wetting, and by washing out the press.

Doubling means that parts of the printed image appear twice, the second time slightly displaced and somewhat more faintly. The printed motif from the previous printing unit is transferred from the paper to the blanket of the next printing unit and is also printed on the next sheet. This happens if the press is not correctly adjusted, or through paper distortion during printing.

Ink stripes occur as a result of the technical structure of the press (→ p. 190, fig. 46). The printing ink is transferred from the **inking unit** to **ink rollers,** which then move the ink on to the print cylinder with the printing plate. In the course of this, the ink passes over several more ink rollers and is squeezed to balance the ink density. Nevertheless, the ink thickness is not absolutely even. The effect shows up in the form of undulating stripes, above all in evenly coloured areas. An attempt is usually made to make the effect less obvious by **adjusting the ink rollers** or changing the water to ink ratio.

There are other effects that can cause stripes: **Channel stripes** are caused by vibration in the printing unit originating in the channel into which the top and bottom of the printing plate are fitted.

If stripes are caused by the **difference drive**, the shear forces often create a sharp edge in the ink at the beginning of a sheet. If motifs present problems in this respect, then hickeys should not be eliminated from the blanket by difference drive.

Tips and Links

Printing

European Color Initiative (ECI):
www.eci.org/eci/en

Fogra Graphic Technology Research
Association: www.fogra.org

Graphics Arts Information Network (GAIN):
www.gain.net

Kipphan, Helmut (Ed.): *Handbook of Print
Media.* Springer Publishing, New York (2004)

Settings and correction

Goldstein, Norm (Ed.): *The Associated Press
Stylebook and Briefing on Media Law.* Fully
Revised and Updated. Perseus Book Group,
New York (2007)

University of Chicago Press Staff (Ed.):
The Chicago Manual of Style. 15th edition.
University of Chicago Press (2003)

Bookbinding

Greenfield, Jane: *ABC of Bookbinding.*
The Lyons Press, Guilford (1998)

Hiller, Helmut: *Dictionary of Books.* French
& European Publications, New York (1991)

Standards

American National Standards Institute
(ANSI): www.ansi.org

British Standards Institution (BSI):
www.bsi-global.com

German Institute for Standardization (DIN):
www2.din.de/index.php?lang=en

International Digital Enterprise Alliance:
www.idealliance.org

International Organization for
Standardization (ISO): www.iso.org

Colour management

European Color Initiative (ECI):
www.eci.org/eci/en

Fraser, Bruce: *Real World Color Management.* Peachpit Press, Berkeley (2004)

Homann, Jan-Peter: *Color Management.*
Springer Publishing, New York (2005)

International Color Consortium (ICC):
www.color.org

Pantone, Inc.: www.pantone.com

Paper

Gmund Paper: www.gmund.com
International Paper:
www.internationalpaper.com
M-real Zanders: www.zanders.com/en
Scheufelen Papierfabrik: www.scheufelen.
com

7.0

What does it take to put a product in the spotlight? What are the ingredients of a great corporate identity? And how do you communicate corporate messages in a range of different media?

According to a well-known truism, "communication is everything". Without a targeted marketing strategy, even the best product risks oblivion. Faced with an ever-increasing glut of information, clever measures that highlight a product's traits and benefits help to tip the scales in favour of its success on the market. To this end, companies tend to pin their hopes on a broad, well-choreographed mix of advertising, sales promotion, public relations and direct marketing.

This chapter introduces marketing's most important visual and psychological strategies; it explains how to communicate a desired claim, how to develop brands and how to use logos, colours, typography and visual language to create a credible, visual identity that sparks plenty of interest in the short-term, yet enjoys a long life.

In view of the sheer wealth and breadth of modern marketing channels (print, web, TV, radio, film, products and even people, building facades or public transport), Marshall McLuhan's famous quote "the medium is the message" has finally lost its edge. Nowadays, the message is all that matters – whether as a classic print ad, internet pop-up, business card or give-away.

Marketing

7.1	**Marketing**	218
7.2	**The Briefing**	219
7.3	**Advertising**	221
	Online advertising	224
	Disciplines in advertising	226
7.4	**E-Marketing**	228
	Search engine marketing	228
7.5	**Direct Marketing**	230
7.6	**Public Relations**	232
	Functions of public relations	233
	Instruments of public relations	233
	The PR concept	236
7.7	**Corporate Identity**	237
7.8	**Tips and Links**	239

Marketing

Marketing today is a complex concept within a massive tangle of interlinked ideas leading to constantly new or revised approaches. These ever-evolving approaches result from inflationary tendencies as well as economic and social change. The term **marketing** was first used in the USA around 1914, where it was equated with sales and understood in a spirit of pure **sales promotion** and orientation. Marketing today offers a systematic and holistic approach, so that decisions can be **directed** both at **markets** and **customers**.

The **American Marketing Association** (AMA) defines marketing as follows: "Marketing is an organizational function and a set of processes for creating, communicating and delivering value to customers and for managing customer relationships in ways that benefit the organization and its stakeholders."[1]

The instruments used in marketing, also called operative marketing measures, are known as the **marketing mix**. The classic marketing mix approach includes product policy (Product), pricing policy (Price), communication policy (Promotion) and distribution policy (Place) – the "4 Ps".

Communication is actually the exchange of information. But the term is so overused today that it is increasingly losing its meaning.

Communication policy – also known as **marketing communication** – is a function of marketing. It includes all the targeted measures aimed at controlling opinions, attitudes, expectations and behaviour to meet a company or organisation's goals. Communication policy is seen as the link between all the instruments in the broader marketing mix. Communication is a permanent process that has to be continuously evaluated – and constantly refreshed – to ensure that it is moving in the desired direction.

Usually, communication policy includes **advertising**, **sales promotion**, **public relations** and **direct marketing**.

[1] American Marketing Association: www.marketingpower.com/mg-dictionary.php

The Briefing

The word **briefing** comes from American military language
and means a deployment discussion with a short description
of the situation and explanation of the aims of the operation,
and a detailed strategy. The term was introduced into advertising by the American advertising executive **Rosser Reeves**[2]
and the advertising copywriter **David Ogilvy**,[3] and was then
adopted by marketing.

The briefing should contain all the necessary information
for conceiving, designing and carrying out a commission.
A good briefing depends on the openness of the briefer and
his or her ability to articulate problems comprehensibly.
Briefings should be set out in writing.

Note: Too much information makes choices difficult, and
thus blurs the distinction between important and unimpor-
tant. The quantity of information should be appropriate to
the commission.

After the first briefing, all the missing information is sought
out and brought together. The **re-briefing** is the subsequent
discussion with the client, an opportunity for corrections and
further clarification after the commission has been acquired.

The so-called **glance over the shoulder** provides a time
for discussion and additional agreements with the client
about the creative approach to the project. This is the stage
at which designers present the basic idea for their concept
to the client, before time and money are invested in the
implementation phase.

2 1910–1984
3 1911–1999

<u>Checklist: Elements of an agency briefing</u>

- Information about the company (e.g. company activities, corporate sector).
- Description of the market situation (starting-point, competitors).
- Commission brief, requirements (e.g. corporate design, tonality) and restrictions, aims.
- Target groups (e.g. core target groups).
- Communication aims (e.g. core messages, positioning, target media, communication mix).
- Product information (qualities, arguments for use, benefits, reason why).
- Budget (e.g. disposability parameters, agency services, other services).
- Deadlines (e.g. launch) and timeline (time and project planning).
- Additional materials.

Advertising

Advertising is a form of mass communication that addresses certain target groups via selected media (**channels of communication**), in order to influence purchasing decisions.

Advertising is always related to trying to attract (**pull**) a certain group of interested persons or potential buyers, while sales and turnover promotion has the primary aim of disposing of or pushing the products available. Advertising should work to the principles of effectiveness, clarity, truth and financial viability.

Ad is short for advertisement. An **advertisement** is a public announcement, commissioned by and issued in the interests of the body making the announcement. It is usually reproduced as a message to be conveyed by advertising, against payment. On the World Wide Web the term "ad" often alludes to **advertising banners** and **advertising pop-ups**.

The basic model for effective advertising

The **AIDA formula** simplifies purchasing processes and represents the ideal achievement of a promotional goal in the form of a model in four stages. This model identifies four phases that the client passes through before deciding to make a purchase. The AIDA formula was developed by the American E. St. Elmo Lewis in 1898 and runs as follows:

1. **(A) Attention** = attracting attention.
2. **(I) Interest** = arousing interest.
3. **(D) Desire** = generating desire.
4. **(A) Action** = bringing about the decision to buy.

According to Richard Geml,[4] advertising operates on different planes. These include **economic aims of advertising** such as sales, profit, profit margin, turnover, sell-off and market share, as well as **non-economic aims of advertising**. Non-economic aims include triggering needs and desires or other emotions, attracting attention and thus creating a positive image.

---

[4] Professor of marketing and economic theory since 1981

Advertising is based on a **sender-recipient model** that sees communication as a process. A formula established in 1948 by the American political and communications theorist Harold D. Lasswell[5] presents a mass communication model. This consists of the **sender**, who dispatches a message to a **recipient** via a certain **channel**, who then responds to the message in his or her turn (→ ch. Design, p. 34).

Lasswell's core questions are: Who (says) What (to) Whom (in) What Channel (with) What Effect?

– Who says …? (sender, communicator),
– What …? (message, content),
– to Whom …? (recipient, target group),
– in What Channel …? (medium, advertising vehicle),
– with What Effect …? (reaction, impact).

Fundamentally, a distinction has to be made between advertising resources and advertising media.

Advertising resources are all the instruments used in sales advertising. Their overall effect is primarily to help meet commercial advertising targets. Thus the advertising resources bundle the messages derived from the advertising target, make them concrete, and present them strongly.

Requirements of advertising resources: originality, succinctness, comprehensibility, manageability, credibility, appropriateness.

The **advertising medium** is the medium chosen to convey the message to consumers. The quality of an advertising medium is assessed by the **cost per thousand** (abbreviated as CPT or, more commonly, CPM): the price per thousand potential interested parties reached.

Advertising media include print media, here above all newspapers and magazines, and electronic media such as the internet, television or radio. Additional advertising media include outdoor advertising, films, products and packaging, promotional gifts, public and private transport, **point-of-sale** advertising (POS) and individuals distributing advertising.

The American advertising expert Rosser Reeves, in formulating the idea that every product (or service) should carry within it a unique sales argument that other products do not share, and that is strong enough to make a sufficient

1902–1978

number of consumers want to buy that product, coined the term **unique selling proposition** (USP). The benefit claimed usually relates to a concrete property of the product or a service that other products (or services) do not offer or do not claim for themselves. The target group(s) addressed in this way should be induced to form preferences for the product (or service) advertised in this way and of course ultimately buy it.

Rosser Reeves distinguishes between a "natural" and a "constructed" USP. A natural USP can be derived directly from a product (or service), its qualities or the way it is produced. A constructed, artificial USP is a claim that cannot be derived from the product itself, its qualities or the way it is produced – and that consequently can only be attributed to it through advertising.

A **testimonial** expresses thanks or makes a recommendation. The phrase testimonial advertising is used for all advertising that is intended to enhance the credibility of the advertising with the aid of statements and opinions from satisfied customers. In the broadest sense, this also includes those forms of advertising that feature celebrities in the role of happy product users or credible spokespersons, proclaiming their satisfaction with the product or service advertised (role model advertising).

Conducting an advertising campaign

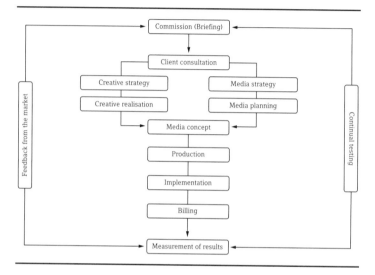

Online advertising

The internet opens up additional pathways for designers
and advertisers with a wide variety of different formats and
technological possibilities. The so-called classical form for
online advertising on the World Wide Web is the **banner**.
Three types are distinguished in terms of function: the static
banner, animated banner or interactive banner.

    Static banners are simple graphic banners. **Interaction**
is triggered by a mouse click, which provides a link to a
particular site or subject.

    Animated banners can be in gif or jpeg format, and can be
presented as a sequence of individual animated images. This
produces a filmic impression of the advertising message, even
though the interaction possibilities are still restricted to a link.

It is also possible to use rich media formats such as HTML (→ ch. Digital Media, p. 136), DHTML, Java, Flash and Shockwave, which support interactive elements such as pull-down menus and selection boxes, and also sound and video effects.

More advanced technologies and the appropriate software have also led to **interactive banners** and other means of communication that enhance the users' attention levels and raise the hit rate. The user no longer needs to leave the website to be interactive. Information, dialogue and action take place within the banner.

Online marketers are working hard on international standards for online advertising forms. As well as the standards listed here, individual marketers offer other formats and special formats internationally.

The following format definitions were fixed by the **Interactive Advertising Bureau** (IAB) (→ p. 239, Tips and Links) in the **USA** as part of their **Universal Ad Package** (UAP):

Format	Pixels	File size for image ad (GIF/JPG)	File size for Flash ad
Super Banner (fig. 1)	728×90	20 K	30 KB
Medium Rectangle (fig. 2)	300×250	20 K	30 KB
Rectangle (fig. 3)	180×150	15 K	20 KB
Wide Skyscraper (fig. 4)	160×600	20 K	30 KB

The following definitions of additional formats are based on the standards of the Bundesverband Digitale Wirtschaft e. V. (BVDW, German Digital Business Association):

Format	Pixels	File size for image ad (GIF/JPG)	File size for Flash ad
Full Banner – also Full Size Banner (fig. 5)	468×60	15 K	20 K
Pop-Up (fig. 6)	200×300	15 K	20 K
Pop-Up	250×250	15 K	20 K
Standard Skyscraper (fig. 7)	120×600	20 K	25 K

The **full banner** is usually placed at the top of a web page. **Pop-ups** present advertising in a special browser window in a way that attracts considerable attention. Such windows open automatically and can be closed by the user – also called **on-site format** (opposite to **in-site format**). Search engines such as Google or Yahoo are committed to more recent **online advertising formats** including innovative use of **mobile phones** or **video**.

Disciplines in advertising

The **creatives** are responsible for the actual design of campaigns, providing the idea and the concept.

As a rule, the **graphic designer** works in tandem with the copywriter. He or she designs layouts, devises presentation charts, inspects fair-copy drawings and printing artwork.

The **art director** (AD) is the artistic director responsible for the visual design of an advertising campaign. As a rule he or she works with a team of copywriters and layout artists; this group also consults the client.

The art director often works with the **creative director** (CD). He or she is the creative head of an advertising agency; he or she co-ordinates the various teams and is ultimately responsible for the creative work. The creative director represents the advertising agency to the client.

The **advertising manager** is the link between agency and client. He or she is responsible for acquiring new clients and is the agency's key contact person here. He or she will work out the briefing details with the client, and pass these on to the creative department. As a rule the advertising manager will also draw up the advertising schedule and make sure that deadlines are met. He or she plans, administers and supervises the advertising budget, directs the client and conveys suggestions and ideas from the creative department to the client.

The **production manager** controls the whole technical production process and makes sure its deadlines are met. The **media planner** chooses the appropriate advertising media. He or she fixes briefings, devises media plans, negotiates discounts and presents the media strategy to the client.

Art buyers buy and organize **free-lance services** for the agency. These include photographers, illustrators, copywriters, draughtsmen and designers. They select suitable specialists and put the working teams together.

The **event manager** devises and organises a whole variety of events.

The **visual media manager** looks after film, radio and television for an advertising agency and is responsible for the products produced there. He or she advises copywriters and graphic artists when they are devising storyboards for film and commercial spots. Once production has been commissioned he or she supervises the quality of the film and sound recordings.

E-Marketing

Search engine marketing

In the USA, Google currently handles 49 per cent and Yahoo 22.5 per cent of all search enquiries.[6] In Germany, Google is ahead with 86.2 per cent of all searches.[7]

To get the best out of search engines it is necessary to know how they work, in order to create the best technical conditions for using them and thus define content design. This is the only way to achieve the best results. The aim of search engine optimisation is to reach the best possible ranking in all search engines. By means of ranking, all search results are placed in order on the basis of search engine algorithms.

Classical search engines have automatic programs that search the **internet** for information. These are called **spiders** or **robots**, with names like **web crawler**, for example. These programs register the content of websites. The spiders read the whole text on a particular internet site. The content is sent back to the search engine and placed in the **database**. The search engine correlates the data on a website that the spiders have sent back.

Tip: *Fundamental point: new and relevant content counts!*

Meta-search engines are characterised by the fact that they do not have a database of their own, but specifically access data from other search engines and web catalogues. A meta search engine does have its own user interface, but dispatches the enquiries to different search engines and web catalogues. The meta search engine also uses its own ranking criteria.

So-called **meta-tags** are information in an HTML document header. These are searched by spiders, robots and crawlers in engine searches and are thus tagged with relevant keywords. Meta-tags do not show in browsers. But search engines do assess some of this information and weight it in different ways. A **keyword** defines a search term according to which the search engines arrange websites in

[6] Nielsen NetRatings study
[7] www.webhits.de

their database. The keywords in their turn are placed in the header of an HTML document. Keywords are words that users type in to define search criteria.

Some search engines assess the popularity and thus the ranking of websites and of individual pages by their link popularity. **Link popularity** is the number of links referring to a particular website. Thus it is also a measure of a site's significance or "weighting". This concept sees each link as a recommendation. The number of external links referring to a website – in other words, the number of "recommendations" it receives – is assessed. The more external links a website has, the more important it becomes, and this also impacts the quantitative aspect in the ranking.

The link popularity concept was developed further by the founders of Google, **Sergy Brin** and **Larry Page**. Google's link popularity algorithm is called **PageRank**. It means that links from websites that have a high link popularity in their own right contribute more to improving a website's ranking, as a qualitative aspect.

Hint: Manipulating link popularity by entries in so-called link lists has now been recognised by search engines and is becoming increasingly ineffective.

A **tracking system**, which periodically provides information about **traffic**, is used to make useful statements about the results and potential of a website in search engines. Uniquely small **tracking codes**, also **pixel counts**, are built into the website in order to work with the tracking providers. Results, analyses or reports can then be called up at any time from the tracking service provider's server.

Sponsored links are paid entries placed on a search engine's results page. Payment is usually on the basis of an auction process in the form of a **pay-per-click** model.

Tip: *Well-known providers are Overture, Espotting and Google.*

Direct Marketing

Direct marketing, also dialogue marketing, is any advertising containing a personal message to the potential client requesting a reply, thus distinguishing itself from classical direct advertising by the **possibility of a response**. Direct marketing has two aims: gaining customers and generating customer loyalty. The latter aims to encourage repeated purchases of a particular product, or to encourage customers to place permanent or long-term orders.

A **response element**, also possibility of **reacting** or **answering**, identifies the possibility of a direct response. A telephone number flashed up during a commercial or featured on a poster is known as a response element, for example. Other response elements are answer cards in adverts and newspaper supplements, coupons, and also **call-me** or **call-back buttons** on a website.

Unlike mass advertising in general, direct advertising can be much more confident in its target groups as it is always personalised. Classical direct advertising includes addressed advertising mailings, direct household advertising such as brochures, catalogues and bulk mailing; partially addressed mailings and also interactive media. Direct marketing in the classical media includes advertisements and inserts with response elements, radio and television marketing with response elements, poster and outdoor marketing with response elements.

In practice, advertising through a mass medium like television or magazines is used to convey an advertising message, and at the same time to offer the recipient the chance to respond. Call centres are a particularly common instrument in this context. It is essential here that the address, telephone number or another response element is part of the advertisement. The aim is to identify anonymous recipients by getting them to express their interest in the company – in other words to respond to the advertisement. Respondents are recorded in a database and are then available for further action, like follow-up offers or reactivation. These measures prepare the way for a direct approach, or make it possible in the first place.

Mailing is the most important element of direct advertising. Replies are recorded in a database, stored, processed, and are then available for future use, e.g. for surveys, reminders, follow-up offers, reactivation (**database marketing**).

The so-called **response quota**, the proportion of positive replies to a communication measure, and **cost per order**, the overall cost of each order or commission achieved, are the major ways of ascertaining whether dialogue marketing has achieved its aims in terms of responses.

Tip: *Simple mailings usually achieve average quotas of between one and four per cent.*

Public Relations

Public relations (PR), also known as **publicity work**, aims to promote and develop trust, understanding and credibility (public image). Public Relations is systematically, strategically planned and methodical communication work and cultivation addressing an organisation's aims and the information the groups it deals with need. PR work inevitably works differently from advertising.

Advertising pays the media for distributing advertising promises, while PR has to persuade journalists above all.

Public relations professionals have a number of tasks to perform. A few of the working areas this mode of communication covers are considered below.

Media relations are the key to PR activities in many organizations. Its aim is to provide print, electronic and online media with information relevant to the company and use them as ways of reaching their target groups. The main media available are newspapers, trade journals, magazines, radio, television and online media.

Public affairs has the job of representing an organisation's interests in the political decision-making process and securing the realisation of its aims. The main target groups are government, parliaments, parties or politicians as disseminators.

Internal communications are an organisation's main means of cultivating relations with its employees and their family members. Tools such as in-house magazines, employee newsletters or intranets are used here.

Product publicity is the term for market-oriented measures to publicise products or services. Because its main target groups are consumers and middlemen, public relations can supply a meaningful complement to classical advertising and marketing communication.

Issues management is PR work in preparation for forming opinions and represents an organisation's efforts to identify public interest in possibly controversial topics at an early stage, to address them and then to help shape the opinion-forming process.

No organisation is immune to crises of confidence, crises in leadership and crises in the financial field, or against scandals or disasters. **Crisis management** involves devising PR strategies and PR measures for the rapid and coherent solution of communication problems in crisis situations.

Financial and **investor relations** secure the company's communications in the equity and financial markets. They are a firm's way of cultivating direct contact with their existing and potential investors. Here the main target groups are shareholders, investors, financial analysts, banks and brokers.

Functions of public relations

1. Contact: conveying information internally (company) and externally (general public).
2. Image: forming, changing and cultivating ideas or opinions about the company (e.g. in relation to people, organisations, specific matters); conveys points of view and makes it possible for a company to find its bearings in terms of creating and securing room for manoeuvre in the process of opinion-forming.
3. Leadership: establishing the positions and decisions of an organisation that set it apart from its competitors.
4. Trust: building up and reinforcing credibility and understanding.
5. Sales promotion: using PR to support sales.
6. Stabilisation: increasing the company's stability in critical situations.
7. Continuity: maintaining a uniform style for the company internally and externally, and in the long term.

Instruments of public relations

1. Press releases and press briefings (news).
2. Face-to-face meetings (press tours, interviews).
3. Press conferences.
4. Background discussions, speeches and statements.
5. User reports.
6. Annual and other market reports.

<u>Online PR</u>

The **internet** is key to today's **public relations**. The degree to which a company and its image are known increasingly depend on how that company communicates with its target groups on the internet.

An open **press section** on the website is usually recommended. This section should meet journalists' needs for information rapidly and keep that information up-to-date. A website can become a more powerful tool for communication through the following factors:

– Rapidly loading pages.
– Clear navigation.
– Consistent layouts.
– Layouts that match the corporate design.
– Text designed appropriately for monitors and target groups.
– A high degree of topicality for content, and easily found information and interaction offers such as press releases, a digest of press clippings, a basic press pack, a calendar of company events and a press contact form (including contact name, telephone number and e-mail address).

The level of acceptance for e-mailed press information has continuously risen in recent years. The advantage lies in less time spent, reduced costs and a greater reach.

<u>Press releases</u>

A press release should be written and published only if there really is something new and important to say. It should also be functional and free of any hint of subjectivity or advertising. The use of unexplained abbreviations, trendy words, superlatives, redundant words or padding should be avoided. The shorter a press release is, the more likely the target group is to read it.

Tip: Provide frank, honest and comprehensive information; limit yourself to essentials. Do not write convoluted sentences, but simple main and subordinate clauses. Use verbal expressions where possible, rather than an undue number of nouns.

Use the active rather than the passive voice. Avoid foreign words if you are not prepared to explain them. Use succinct but not sensational headings and subheadings. Match the photographs to the text.

Half to one page of A4 is sufficient for a press release; on no account should it be longer that two pages of A4 (→ ch. Production, pp. 152 – 155, paper formats). The lines should not be too closely spaced and the text should be justified to the left margin.

Tip: Use 1.5 line-spacing so that journalists can write notes and leave a wide right-hand margin for further notes and comments.

Paragraphs are separated by a blank line. Font and point size should be clearly legible. The logo and the sender details should be on the first page. Direct speech should be clearly identified as quotation. First name and surname should be given in full, with title and position where applicable.

<u>Structuring a press release</u>

Fundamentally every press release must answer the following six **W questions**: **who** said **what**, **when**, **where** and in **what way** to **whom** (or what happened to whom, when, where, in what way and why?).

1. **Headline:** Slogan-like, making a key statement.
2. **Subhead:** More brief information.
3. **Lead** – also **introduction, teaser:** The first five lines, at the most, must sum up the essential content of the message; if possible, all the W questions should be briefly addressed here.
4. **Central section:** Information, explanations, details, background and detailed response to the W questions; it may be structured as individual paragraphs with subheads. This section is constructed on the basis of the inverted pyramid. This means the most important items come first, and the least important ones towards the end of the text. Then editors can cut the text by starting at the end.

5. **Boilerplate:** Background and additional information about the company, summary of the company's activities.
6. **Contact address:** Contact person with postal address, telephone, fax, e-mail and internet address.
7. **Other information:** E.g. downloading instructions.

Tip: *Don't forget the date and place in the lead!*

The PR concept

Communication planning is not the only place where the concept is the key element. The **PR** or **communication concept** is a methodically developed and lucidly structured planning paper. Its length depends on the concept type and task in hand.

Checklist: PR concept content

– Analysis of actual vs. target situation.
– Analysis of strengths-weaknesses, opportunities-threats (SWOT analysis).
– Definition of communication tasks and aims.
– Definition and profiling of target groups.
– Formulation of key messages.
– Formulation of communication strategy.
– Description of instruments and communication measures.
– Timing, costs and organisation planning.
– Evaluation.

Corporate Identity

Corporate identity is now a strategic instrument used very purposefully in **company management**, and contributes to the success or failure of an enterprise. Corporate identity (CI) defines a company's personality, character and attitude. It addresses the way in which the company perceives itself, and also its ability to communicate its own aims internally and externally. These guidelines can be summed up in a **CI manual**.

An organisation is able to present itself credibly and authentically only if its self-image and the impression it makes on outsiders are in harmony. The following characteristics are key to an effective corporate identity:

1. Unmistakable and striking qualities.
2. Succinctness.
3. Comprehensiveness and consistency.
4. Credibility.
5. Continuity and reliability.

Coherent and strategic use of all communication techniques can express an individual, uniform company identity that is free of contradictions. The classical approach represents the corporate image and its components as follows (fig. 8):[8]

---

[8]

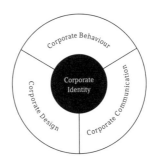

---

[8] From Birkigt, Klaus; Stadler, Marinus M.: Corporate Identity (2002). Not available in English.

<u>Corporate identity phases</u>

1. The identity process.
2. Market development.
3. Communications strategy.
4. Design development.
5. Migration and documentation.
6. Checking the measures taken (monitoring).

Corporate design brings the complex qualities and achievements of a corporate personality together and forges them into a memorable, uniform and credible visual statement. When a company's corporate design matches its corporate identity, then its statements on identity and appearance will be credible.

The **basic elements** contributing to a corporate design's unique overall impact are the **logo** (→ ch. Design, pp. 37–40), the **colour climate** (e.g. corporate colours), **typography** (e.g. corporate fonts), the **pictorial language** and the **design principles**. The basic elements are fixed in writing in a **CD manual**, also called a **style guide**.

Tips and Links

Online advertising, e-marketing and online PR

Google Webmaster Help Center:
www.google.com/support/webmasters

Interactive Advertising Bureau (IAB):
www.iab.net

Nielsen NetRatings:
www.nielsen-netratings.com

Marketing

Kotler, Philip; Armstrong, Gary: *Principles of Marketing (11th Edition)*. Prentice Hall, Upper Saddle River (2005)

Advertising

Art Directors Club: www.adcglobal.org

Lürzer's Archive: www.luerzersarchive.com

Associations and organisations

American Marketing Association (AMA):
www.marketingpower.com

International Public Relations Association
(IPRA): www.ipra.org

Public Relations Society of America (PRSA):
www.prsa.org

8.0

A professional designer is aware of the laws and regulations that affect and protect his or her work.

That's the theory anyway. In practice, most designers prefer to ignore the details of legal issues. Armed with only a vague inkling of their basic rights and the rules that affect them, they quickly find themselves out of their depth when an actual legal issue arises. Where exactly are the boundaries between legitimate homage and blatant plagiarism? Do websites enjoy similar copyright protection as, say, photographs or oil paintings? What happens in case of a breach of contract? And what about the bigger picture of the idiosyncrasies of international legal systems?

Faced with the intricacies of national and international legislation on copyright, design patents, contract law, payment modalities, liability, protection of intellectual properties or rights of use, even the most experienced design professionals often find themselves stumped. Globalisation, for example, which is a fount of potential new business opportunities, comes with just as many legal ambiguities; these can quickly turn into sizeable stumbling blocks when we leave the cosy confines of our national territory.

The following pages aim to shed some light on the current state of international law and offer some vital hints on how to avoid most pitfalls, disappointments and risks in advance. Knowing your basic rights and duties, which differ from those of a fine artist, can help you smooth out any bumps in the road – even before you embark on your project.

Law

8.1	Important Note from the Legal Translator	244
8.2	Contracts and Commissioned Orders	245
8.3	Design Protection	254
8.4	General Laws of Protection	255
	Copyright Law	255
8.5	Protection Rights for Industrial Property	270
	Registered Design Law and Design Patent Law	270
	Trademark Rights	272
	Competition Law	275
8.6	Tips and Links	279

Important Note from the Legal Translator

This section provides an overview of the legal matters that are relevant to designers. It is intended for general guidance only. Furthermore, this overview has been made on the basis of German law, with the result that some of the legal concepts set out herein may not be applicable (nor even existent) in your legal system. Consequently, it is recommended – before taking any legal action – that you take legal advice from a lawyer with experience in media industry matters, while at the same time being duly authorised to advise you with regard to legal matters relevant to your applicable jurisdiction (and therefore to the applicable law).

Subject to the foregoing, however, it must also be said that German law is very detailed and thorough, and therefore well suited to give a general overview of legal matters which are relevant to designers. Where possible, we have also given a few examples of how certain legal matters are treated under non-German legal systems (i.e. most notably under Anglo-Saxon legal systems), never forgetting that law, as a human construct, always maintains a complex, national identity.

Contracts and Commissioned Orders

<u>What should the contents of a legal contract look like?</u>

The most important prerequisites for successful collaboration between designer and client are clear terms of agreement, as well as open communication and transparency. Whosoever – subsequent to a verbal agreement – attempts to submit a substantially divergent offer to the client, or announces additional business conditions after the fact, which furthermore contain unacceptable supplementary elements/conditions, engenders mistrust, detracts from a productive working atmosphere, and jeopardises further collaboration.

Contracts provide the legal foundation for every single exchange and transaction of and/or for services. Whoever offers or tenders their services to another person, and receives payment for same, has in effect made a contract. It is advisable to render contracts in written form, as a general rule, although verbal agreements have equal validity. Nevertheless, with verbally negotiated contracts, it might be difficult, or eventually impossible to prove the original points of agreement, or even the original placement of the order (i.e. at a later point/date in time).

To the extent a designer is obliged to settle/conclude a diverse number of contracts with similar content, it is advisable to use standard terminology, which can be utilised for a variety of contractual agreements. The so-called **General Terms and Conditions of Business** (GTCB) are standardised terms of contract rendered in written form. It suffices when – in the course of negotiating commissioned work – designers refer to their GTCB while attaching these to the actual contract or to the reverse side of the offer submitted to the potential client. These terms of business are then legally binding, i.e. to the extent the client has not raised objection.

Tip: *The following text can be attached to the offer of services, or to the official confirmation of services submitted (in clear and written form!) to the potential client or commissioning party: "The contractual conditions attached to the reverse side of this document form part of the agreement and are legally binding."*

The offer of services

Initially, the foundation of every contract is the offer of services, which is generally preceded by a verbal agreement with the potential client. The official offer should contain all points essential to the commissioned work. If the description of tendered services proves to be sufficiently comprehensive, the respective conceptions of the persons involved will be correspondingly clear. The most important aspects and contents of a service offer are listed below:

1. **Description of services.** Which services should the designer provide?
   For example, the compilation of a 32-page exhibition catalogue, corporate design documentation in the form of a style guide/CD manual, or a concept for business stationery encompassing writing paper, business cards and envelope designs.

2. **Third party services.** Which services should be delegated or subcontracted by the designer?

Tip: *Due to high* **liability risks**, *the designer should not subcontract commissioned work to third party businesses such as printing presses, among other possibilities. Should the designer assume responsibility for the supervision of printing services, however, it must be documented in written form that the applicable service order will be initiated and placed directly by the client in collaboration with the printing press, or that the designer will supervise the service order on behalf of the client, i.e. per procurationem (p.p.). In the latter case, the designer should obtain valid authorisation to commission the required publishing services directly from the client. The following text excerpt can be adapted to the given case: "To the extent the designer must of necessity order third party services, the latter commissioned parties are not in any legal sense auxiliary persons of the designer. The designer is exclusively liable for the consequences of his/her own actions as regards premeditation and gross negligence."*

3. **Payment for services.** Which payment must the client make to the designer?
   The designer receives an appropriate payment for services rendered. The latter is to be defined in the form of a cost estimate contained within the service offer. As a matter of principle, expenses should never be "concealed" and should not appear in the invoice for the very first time, as this can only lead to understandable consternation on the part of the client.

Tip: The following formulation can be used for this purpose: "The client pays the designer the following drafting fee XY for the production of YZ."

The designer can negotiate either an **hourly wage** or a **flat-rate/fee payment**. **Extraneous costs** or **further expenses** – so-called "incidental technical expenses" – can be, for example, messenger deliveries, model fees or proof processing expenses (→ p. 252). All extraneous costs or further expenses are to be listed separately in the service offer.

Tip: To the extent all expenses are not foreseeable, it is also possible to negotiate that the client assume all responsibility to remunerate extraneous costs. These should then be submitted by the commissioned party to the client without any extra charge. The following text excerpt can be adapted to the given case: "The client or commissioning party has to reimburse the commissioned party for all extraneous costs actually incurred."

In the service offer, a reference should be made to the necessity of calculating and including the net amount paid with the statutory value-added tax (VAT).

Tip: The legal interpretation of contracts occurs from the perspective of the "recipient's horizon", i.e. "unclarities/ambiguities are judged to be the fault of the tenderer (i.e. the party producing the document)."

Designers in the **Federal Republic of Germany** (→ p. 244, Note from the Legal Translator) usually work with their respective clients on the basis of **contracts for labour and services** (in German: Werkvertraege). In such cases, the completed production of the contractual work is due; whereas under service contracts, the rendering of a particular service is pending.

According to **German labour and service contract law** (→ p. 244, Note from the Legal Translator), payment is due only subsequent to the **acceptance** (→ pp. 250, 251) of the correctly produced contractual work; or, under certain conditions, subsequent to the acceptance of autonomous parts of the whole contractual work. Therefore, precise **terms of payment** should be agreed upon. Along with concrete agreements regarding the **type** and **volume** of the commissioned work, as well as the **amount of payment**, it is also essential that **payment deadlines** be set.

Tip: *In the case of voluminous or temporally extensive projects, it is recommendable to negotiate partial payments, and thus precisely determine which payments will be due at what time.*

4. **Exploitation rights and rights of use.** Which rights are to be assigned to the client by the designer? The assignment of rights of use (→ pp. 263–266) should be negotiated on a contractual basis and already presented in a transparent manner within the service offer. This means: Rights of use should already be a settled matter during the actual process of officially commissioning the required services.

Golden Rule: All expenditures listed in the invoice should have been contained within the original service offer. *In no event should it be presumed that the clients or commissioned parties will remunerate what the designer has charged to them without prior consensus.*

The **German Civil Code** (§ 632) (→ p. 244, Note from the Legal Translator) does indeed provide for the legal validity of a payment in the form of a "tacit agreement", "if under the specific circumstances the production of the work can only be expected in return for payment". Notwithstanding, a clear and transparently formulated agreement is always preferable.

5. **Corrections by authors.** The agreed remuneration should not cover more than three corrections by authors for the contractual object. The number of additional corrective phases as well as their respective hourly wage should be displayed separately within the service offer.

Legal rule of thumb

As a **general rule** it holds that: When in doubt in respect to legal stipulations for the purposes of contractual determinations, for example where the designer and the client are based in different countries, the designer should provide that the law of the designer's country should apply and that the courts of the designer's country should deal with any disputes.

The following text passages are admissible for this purpose: "The place of execution is the main business office of the designer"; and "The legal provisions of XY (country) are valid and applicable in this context, and under this agreement."

Thus, designers located in the **Federal Republic of Germany**, for example, have the advantage of access to extensively articulated legal provisions, whereas in the **United States of America**, by contrast, the subsumption of such legal considerations has been predominately relegated to **case law**. As a rule, designers operating within the USA are well advised to render all substantial points and aspects thoroughly in written form.

Commissioning Work

If an order has been placed, or if work has been commissioned, the designer typically commits him/herself to the production or creation of the contractual work. Usually, it already suffices if the client or commissioning party has signed the service offer.

Tip: It is advisable to preserve the entirety of written agreements subsequent to contractual settlements with a given client. Verbal agreements should be rendered in the form of a protocol, after which the client should be requested to provide official confirmation of the document; this may be carried out by means of an e-mail. Should a dispute develop with the client in respect to the contracted services, the designer must be in a position to concretely verify – and in extreme cases even prove – what exactly had been negotiated with the client up until that point in time in respect to the contracted services.

Acceptance

Acceptance is essentially the receipt and recognition of **services as a contractual act**, i.e. through the corresponding affirmation of the client or commissioning party.

According to **German labour and service contract law** (→ p. 244, Note from the Legal Translator), provided no special agreements have been made, payment is due only subsequent to the acceptance of the contractual work. The commissioned party with its agreed service performance is considered to be "in arrears" in respect to the client, essentially owing the client a finished defect free result, so that the commissioned party will have first fulfilled the terms of contract once the service/performance rendered services do not have and/or contain any significant deficiencies/defects.

Tip: The official acceptance should be documented in written form in order to allow the transaction to be substantiated at any time. It is advisable to draw up an acceptance protocol, which could contain passages similar to the following: "The client has received the following services …" (if need be, with corresponding illustrations attached). "The following defects have been discovered …" with location, date and signature of the client. Additionally, the handwritten signature of the client on a draft is also suitable, e.g. with the following note: "The performance has been confirmed to be in due fulfilment of the contract."

Note: An e-mail is only then equally valid evidence, when an electronic signature has been utilised, which is an extremely rare occurrence in the daily practice of a designer. In the absence of an electronic signature, the commissioning party could always contest the fact that the e-mail had originated from its side.

Confirmation with the signature of the client and by means of a fax is the more reliable method.

It could be the case that the client or commissioning party objects to payment claims raised by the designer (as commissioned party) as a result of claimed defects in the agreed contractual performance, in which case the invoice sent by the designer remains unpaid.

As regards **German law** (→ p. 244, Note from the Legal Translator), the following must be heeded:

1. Subsequent to the acceptance of goods or services, the burden of proof in respect to defects is transferred to the commissioning party;

2. and subsequent to the acceptance of the work, the commissioning party may only enforce warranty rights for defects identified (by it) at the time of its acceptance of the contractual work, provided the commissioning party has reserved such rights for itself during the acceptance phase of the contracted goods or services.

The invoice

As a **general rule**: The content and items contained within an invoice should correspond to the content and items contained within the original service offer. In order to facilitate auditing procedures for the client, the original service offer should also be submitted along with every invoice.

1. Furthermore, an invoice should include the complete **name** and **address** of the **party providing services** (designer, accountant for the invoice), as well as the **bank account** used for business transactions and the corresponding bank data. For designers in the Federal Republic of Germany (→ p. 244, Note from the Legal Translator): In order for the invoice to entitle its recipient to deduct the itemised input tax (this law presumes actual residency and not only registration at the Citizens' Registration Office), besides the information regarding name and address indicated above, a valid invoice must also contain the following information (Value-Added Tax Law § 14, Para. 4): The **tax number** or **value-added tax identification number**, the net remunerative sum itemised according to the respective tax brackets, and the taxed component of the remunerative sum.

2. **Extraneous costs** and expenditures must be itemised separately; however, these must have been agreed upon beforehand (→ p. 247).

Tip: *If the precise fiscal value of extraneous costs cannot be established, the reasons for these expenditures must be named, such as travel costs, technical expenditures etc.*

3. **Payment deadline.** In the invoice text, the **time period allowed for payment** – as well as a **payment deadline** – should be indicated. According to German service contract law (→ p. 244, Note from the Legal Translator), payment is **immediately** due subsequent to the acceptance of the contractual work. The payment deadline may therefore be set within a relatively brief period of time.

Tip: *The payment deadline should be determinable in correspondence to the standard calendar. Even better, however, is the determination of a specific date or period of time such as "payable within 14 days", for example.*

<u>What if the client doesn't pay?</u>

Tip: The time span between the receipt of the invoice and the first reminder should not be all too long. Two to four weeks after the designated payment deadline is appropriate.

To the extent the client or commissioning party has not paid within one month after receipt of the invoice, given that no other allowable time periods for payment are indicated in the GTCB of the commissioning party, the latter is subsequently in arrears and is obliged to compensate the designer for **damages caused by delay**. Nevertheless, if any doubts arise, the designer must be in a position to prove the successful delivery of the invoice.

The **damage** – within the context of defaulted payments – is the interest which the designer must pay to his/her bank, provided this can be substantiated, whereas the generally accepted interest rate for default without such evidence would be 5 per cent above the standard interest rate (i.e. the recommendation according to German law). Moreover, costs for legal counsel (i.e. lawyers) may also be charged, provided the designer has enlisted legal counsel in order to prepare payment notices, and that same have been sent by an applicable lawyer. Under certain circumstances, the designer may also acquire the right to refrain from continuing the contracted service.

Tip: Various calculation models, payment tables, contractual aids, as well as other model texts and legal counsel are available through professional and trade associations.

Design Protection

As already stated in the note from the legal translator, it is nearly impossible to furnish a precise classification and allocation of design protection laws in the various spheres of international jurisdiction – not least because of the highly divergent premises at the very core of the issue as to which purposes industrial design should fulfil, since such premises are subject to continual development.

The principal questions are: Which criteria should provide the foundation for design protection, e.g. the aesthetic effect, or rather, the functionality of a product? Should the commercial aspects, the creative process or the immaterial, i.e. intellectual commodity be protected?

Thus, according to the given country and corresponding to varying criteria of evaluation, the form and function of a design are considered from many different perspectives, and thus, varying prerequisites have been determined as to what defines protectability. Principally, a patent is intended to protect the function of a given product, whereas the copyright protects the creative powers of expression, and trademark laws govern the ability of commercial labels to distinguish between different originators. In turn, registered design law is usually assigned to the jurisdiction of copyright law.

General Laws of Protection

Copyright law

Copyright law was established in order to support and secure royalties for creative accomplishments as well as "artistic innovations".

In principle, it holds that: The prerequisite for the copyright protection of a design is that a work has been created with a performance in design transcending/surpassing the overall average. There exists no internationally valid copyright protection. Authorship protection is dependent upon the legal regulations of the respective countries. There exists the so-called **territoriality principle**. In effect, protective regulations are limited to the territory of a given country.

In addition to this, agreements and contracts such as the **Revised Berne Convention** (RBC) or the **World Intellectual Property Organisation** (WIPO) ($\to$ p. 279) exist, which set internationally valid and binding minimal standards for each national jurisdiction.

The **Universal Copyright Convention** (UCC) ($\to$ p. 279) also protects creative productions in countries not belonging to the RBC, or countries that were unable to meet RBC standards. The UCC defines and protects minimal rights of translation, reproduction, presentation, recitation, broadcasting, as well as processing. The Universal Copyright Convention was established in 1952 through the initiative of UNESCO, in order to allow such states as the USA, which function with an entirely different legal system, to have access to a legal accord of broad international scope. In 1994, 94 states had signed the agreement. The USA joined the RBC in 1989.

Generally speaking, two distinct international copyright systems have evolved: On the one hand, a copyright system operating on immaterial criteria and characterised by "artistic leanings", such as German copyright law, on the other hand, a copyright system orientated to economic principles and commercially influenced, such as the US copyright system ($\to$ pp. 258–262).

The continental copyright system

The **continental copyright system** holds the originator and his/her personal rights as a creative individual to be the central factors. An example of this emphasis is copyright law in the **Federal Republic of Germany**.

German copyright law protects the originator on the basis of his/her personal rights: "The originator is the creator of the work" (German Copyright Law § 7). Hence, the German copyright is – so defined – not transferable and can only be inherited, or, so-called rights of use can be negotiated for the creative product (→ pp. 263–265, concession of rights of use).

German copyright protection ensues **automatically** with the **creation** of the work, which makes registration for this purpose both unnecessary, and in principle, impossible. German law defines the "**work**" as a personal intellectual creation (German Copyright Law § 2, Para. 2). The necessary condition for the latter is that the work must evince an independent artistic character, a certain degree of individuality, as well as an intellectual content, the so-called creative level.

German copyright law protects works of literature, science and art (Protected Works § 2). Among these – in accordance with copyright law – are, for example: "linguistic productions such as written texts, speeches and computer programmes"; "musical works"; "works of the fine arts, including architectural art, applied art as well as the drafts and designs of such productions"; "photographic works as well as all such works produced with similar means"; "works of film as well as all such works produced with similar means"; "illustrations of a scientific or technical nature such as drawings, designs, maps, sketches, tables and graphic portrayals".

Note: With respect to **German** and **Italian art**, the threshold to warranty protection is extremely high, i.e. how much creative accomplishment and creativity a work of applied art must demonstrate in order to be granted copyright protection (→ p. 270, formative level).

Practical experience proves that an overwhelming number of creative productions are never recognised to be deserving of copyright protection. Within this context, one distinguishes between the **fine arts**, which are nearly always granted copyright protection, and **applied art**. According to

German law, design is classified in the latter category, which in turn rarely receives copyright protection.

In the event that the creative production cannot be protected in accordance with copyright law, a so-called **pattern design** can be registered. Already in the registration phase of pattern design, the German Registered Design Act requires that both the design and its realisation be distinguishable from an average production, from pure "handiwork" and from everyday objects (→ pp. 257, 270, pattern design).

Note: Even if a designer has created a new work on the computer screen by manipulating an image, for example, that original prototype is still protected under copyright law; unless it happens to be the case that the usage is judged to be free as regards copyright law, meaning, in effect, the "old" work had been "eclipsed" by the new work. Even if this is the case, such matters are extremely difficult to clarify. It is advisable to acquire the permission of the originator in advance.

Copyright protection in Europe. The European Union has issued numerous regulations and guidelines to create a general copyright norm for its member states.

Thus, Directive 91/250/EEC from 1991, for example, protects **computer programs** as "works" in the legal sense of copyright law. In 1993, the **period of protection** for those **works** of literature and art entitled to copyright protection was fixed uniformly at up to 70 years following the death of the originator. Thus, the EU had at the same time created the foundation for a Europe-wide, uniform incorporation of its contracts into national jurisdiction. European legal provisions concerning copyright law have been put forth in EU copyright regulation (Directive 2001/29/EC).

Note: A website as such – wholly independent of its particular content – enjoys legal protection as a computer programme. Further, should an individual internet homepage happen to be linked to a database, that database can also be protected in a like manner. For the legal protection of a database, together with the corresponding recognition of its personal intellectual creation, it already suffices that the respective data units have been processed with a certain expenditure, and thus, that the criteria for personal intellectual creation have been met.

French copyright law is put forth in the "Code de la propriété intellectuelle" dating from 1 July 1992. In the case of French copyright law, as well, there exists – due to the nature of the matter – an immaterial right possessed by the originator in respect to his/her creative product (creation principle).

The Anglo-Saxon copyright system

American copyright was first codified under the Copyright Act of 1790 and is essentially based on the "Statute of Anne", the first English copyright law from 1710. American copyright was revised in 1909 and adapted to changing social circumstances, though this legislation still fell well below the global standards of the Berne Convention, concluded in 1886 and revised in 1896. For example, the US Copyright Act of 1909 continued to differ from the standards set by the international agreement in that it continued to state that an author had to comply with certain formalities in order to enjoy copyright at all.

Copyright protection did not exist automatically once the work was created. Copyright had to be registered at the Copyright Office established as a department of the Library of Congress in 1897, and then and only then could intellectual property be asserted. Copyright also continued for considerably fewer years (limited to 28 years), even though the protection could be extended for the same number of additional years. But certain conditions also had to be met in order to extend copyright protection, and the extension had also to be registered. The 1909 Copyright Act was adapted in some respects in the course of time, so that unpublished works could also be protected, for example (until this time unpublished works were at best protected by parallel laws in individual states).

But it was precisely against this background of the lack of international standards in American copyright that the USA passed a copyright amendment in 1976, which became law in 1978 (for individual regulations see www.copyright.gov/title17/). Finally, after further legislative changes to the Copyright Act, the USA became the last industrialised nation to adopt the Revised Berne Convention on 1 March 1989.

Despite this, US copyright still fundamentally works on the basis that investment in the production of a work is the key. It is immaterial here whether this investment is made by a company or an individual. Continental understanding of copyright is still more immediately concerned with a creative act by an individual, seeing that individual as creating work that is thus worthy of protection; but American copyright law placed commercial exploitation of a work in the foreground. So American copyright law, which like German copyright law protects various genres of work (17 U.S.C. § 102), can also be transferred in full to third parties (17 U.S.C. § 201).

The "**works made for hire**" doctrine (17 U.S.C. § 201) represents a fundamental difference from German copyright law and confirms the essentially commercial view taken by American copyright law.

A work made for hire is defined as commissioned work that is either created by an employee within the scope of his or her employment, or that was specially ordered for a specific purpose. But both parties must expressly agree in a written instrument signed by them that the work shall be considered a work made for hire. Then, in the spirit of American copyright law's essentially commercial view, the law logically assigns the original copyright in the work to the employer or other person for whom the work was prepared (17 U.S.C. § 201).

Hence a contractor, if a commission is subject to American copyright law, becomes the author of all work created by the employee at the point the work comes into being, without any use or exploitation rights having to be transferred. This does not apply automatically, however, but only if the conditions for a work made for hire are met, in other words if an appropriate written instrument is signed by employee and employer, in which the work is expressly defined as a "work made for hire". Verbal agreements are not sufficient.

German copyright law is planning a ruling related to the work made for hire doctrine in the sphere of software programming. § 69 b of the German Copyright Law applies if authors create work on the basis of working relationships or relationships in contractual law. Here, too, the employer is directly and exclusively entitled to disposition over software created by the employee. Use and exploitation rights do not have to be transferred to the employer, provided that the software has been created under appropriate contractual conditions or on the employer's instructions. Despite this, under German law the software programmer remains author of the computer program in question, and thus owns the moral rights (in German: "Urheberpersönlichkeitsrechte"), even though that person cannot exploit his or her work personally at any time.

But a contractor or employer who becomes author of a work made for hire under US law can exploit this in the same way as the owner of the copyright or the proprietor of an economic asset and freely transfer the worked commissioned in whole or in part to third parties. Thus all rights of commercial exploitation are withdrawn from the actual "creator of the work". The only compensation the American creator has for his or her work is a claim to remuneration as per contract or to payment for commissioned work. He or she cannot, however, claim rights in the work beyond this or part of the profit from subsequent exploitation of the work, possibly in various media.

In commercial life, this confers considerable legal relief on the producer of a large-scale work involving several persons making contributions subject to copyright protection (for example in the production of a film). While under German law outside the concrete purpose of the commission all rights of use and exploitation are transferred individually from the author to the employer, and could possibly have to be agreed additionally at a later time by each author involved (against additional remuneration), under US law the employer can secure the copyright for him- or herself with a single commissioning instrument, without having to face the fact that he or she can be subject to claims from an employee at a later date.

Registration is still recommended, even though giving notice of, or registering, copyright in foreign works (i.e. works that were first published outside the USA) with the Library of Congress is no longer compulsory in order to pursue infringements of the law under the 1976 Copyright Act except in certain special cases. If the work is registered within five years of its publication, the author acquires enhanced evidentiary weight in the question of copyright ownership (so-called "prima facie evidence of the validity of the copyright", 17 U.S.C. § 410). Additionally, registration establishes a public record of the copyright claim, and that someone has acknowledged that the work has been created (relevant in the case of a dispute about who is the original author of the work). In addition, US courts grant statutory damages and legal costs only in the case of registered works (17 U.S.C. § 412). Given that legal costs are so high in the USA, this can be highly significant in terms of whether the author can pursue infringements or enjoy copyright protection.

In order to register with the Copyright Office, the author must submit, in an envelope, a properly completed application form; pay a single, non-returnable filing fee and provide a voucher copy for the Copyright Office (17 U.S.C. § § 408, 409, 708).

Another important innovation in the 1976 Copyright Act is that it extended the protection afforded under copyright for all works created after 1 January 1978. In order to meet the minimum requirements for admission to the Revised Berne Convention the USA granted protection in the author's lifetime plus 50 years after his or her death. As of 27 October 1998, an amendment ("Sonny Bono Copyright Term Extension Act") extended copyright protection to 70 years after the author's death.

The **copyright notice** formality also used to be of central importance in the USA. It was only by identifying the date the work was published through the copyright notice that the (short) protection term could begin. In addition, copyright could lapse into the public domain (making the work available for general use) if the work did not have a copyright notice.

As the modernised 1976 Copyright Act introduced linking the copyright term to the author's death, the copyright notice has been irrelevant for calculating the copyright term since January 1978. In addition, since 1 March 1989, the date the USA adopted the Revised Berne Convention, the copyright notice is no longer compulsory in the USA.

Hence the author can no longer lose copyright simply because his or her work does not carry a copyright notice.

So now – with a very few exceptions – the copyright notice serves only to provide public information.

Nevertheless, it is still advisable to use the copyright notice to identify the work and as an indication that an author created the work at a given date. An indication of this kind can be given by the symbol (fig. 1), the word "copyright" or the abbreviation "copr.", in each case with the year of publication and the author's name (17 U.S.C. § 401), for example "© 2007 Die Gestalten Verlag GmbH & Co KG". In the case of sound recordings, a "p" is placed in the circle instead of a "c" ("phonorecord"), and next to it the year of first publication and the name of the copyright owner (17 U.S.C. § 402).

As well as providing information for the public, applying a copyright notice to a work or sound recording also has legal significance. In legal proceedings about copyright infringement the accused party cannot object that he or she committed the infringement accidentally. The damages due to the author cannot then be reduced, while otherwise – in the case of an accidental infringement – the court would have to reduce damages where appropriate.

The UK Law of Copyright is valid in England, Wales, Scotland and Northern Ireland. The general approach taken under UK copyright law is similar to that taken under US copyright law rather than that of copyright law in other European countries. Copyrighted works are regarded primarily as commercial products rather than artistic creations. The authors and creators of those copyrighted works are allowed to transfer the copyright freely on whatever terms they wish. Copyright is a form of property right protecting certain types of human creation, which are recorded in some form.

The general position is that the duration of copyright protection in Great Britain for works created since 1996 expires 70 years subsequent to the death of the author or creator.

<u>Infringement of copyright</u>

In the event of an infringement of copyright, for example by unauthorised copying, the copyright owner, is entitled to an injunction (a court order preventing further infringement), compensatory damages or, alternatively, payment of the profits made by the infringing party and orders for the delivery of infringing copies and their destruction.

<u>Exploitation rights</u>

The **concession of rights of use** is the most important instrument for the **exploitation** of **copyright**.

German copyright – exploitation rights (German Copyright Law, General Points § 15):

1. The originator possesses the exclusive right to exploit his/her work in its physical form, whereby this right encompasses the right of reproduction (§ 16), distribution rights (§ 17) and exhibition rights (§ 18).

2. Furthermore, the originator possesses the exclusive right to reproduce his/her work in a non-material form (right of public reproduction). The right of public reproduction encompasses, most importantly, the rights of recitation, performance and presentation (§ 19), the right to public accessibility (§ 19 a), broadcasting rights (§ 20), the right of reproduction by means of visual or auditory media (§ 21), the right to reproduce radio programmes and make these publicly accessible (§ 22).

3. Reproduction is to be considered "public" if it has been determined for several members of the general populace. A member of the general populace is any individual who is not – by means of personal relationship – in any way connected with the subject exploiting the given work, or with any other persons for whom the given work has been made accessible or perceivable in a non-material form.

Concession of rights of use (in accordance with German Copyright Law § 31):

The originator may concede the right to make use of the work for specific uses, or for all types of usage, to another party (rights of use). These rights of use may be granted on a basic/simple or exclusive basis, may be conceded territorially or temporally, or may be made contingent upon restrictive stipulations. "**Basic**" or "**exclusive**" characterise the legal nature of utility right concessions, whereas "**conditional**" or "**unconditional**" describe the extent of utility right concessions, such as determining whether a work may be utilised regionally, or utilised with limitations either connected with time considerations, or directed at particular contents of the work.

Should the designer concede **basic rights of use**, he/she may then continue to make use of the work and to concede basic rights of use to further individuals (fig. 2). The holder of the rights of use (licencee) cannot for his/her part concede any rights of use to third parties, unless such an arrangement has been expressly negotiated.

2

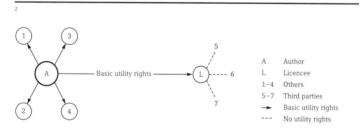

A	Author
L	Licencee
1–4	Others
5–7	Third parties
➞	Basic utility rights
---	No utility rights

Exclusive rights of use are an **exclusive right** as well as the most frequent form of utility authorisation delegated by designers, as it is usually undesirable to the client or commissioning party that others make use of or exploit the same design. In respect to exclusive rights of use, the originator of the creative product may not concede any rights of use to third parties, and may not make use of the work him/herself.

The holder of the rights of use may in turn – with the approval of the originator – concede basic rights of use to third parties (fig. 3).

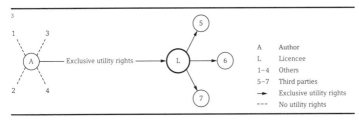

Broadly viewed, the concession of exclusive rights of use corresponds to the logical consequence of the "work made for hire" principle in the USA. The commissioned party in the Federal Republic of Germany, however, always remains the originator of the work he/she has produced, although at the same time, the commissioned party may be excluded from any and every usage of the same creative product. It is also possible to agree upon exclusivity for merely one specific area, e.g. for the internet. Accordingly, the designer would then be authorised to exploit rights of use in other areas such as printed media, etc., or to make personal use of the work. In the particular case of German law – in contrast to jurisprudence in the USA and Canada – it is not possible to concede the entire copyright as a whole.

US copyright law – copyright concession/assignment.
As copyright is an economic asset in the eyes of copyright law in the USA, and the creator of the work is not necessarily entitled to it, and it can be sold in its entirety, it is possible on the basis of copyright law as valid in the USA to transfer copyright completely to third parties, by selling it, for example. Selling or otherwise disposing of copyright should not be confused with the transfer of use and exploitation rights covered in continental European copyright legislation, even though the right to enjoy rights of use in a work and to exploit them commercially may produce a similar commercial result.

According to the 1976 Copyright Act, transferring copyright must be undertaken in writing and at the same time as the agreement to legal transfer (17 U.S.C. § 204). In addition, when copyright or exclusive rights in it are transferred, the owner of the copyright or such owner's duly authorised agent must sign the written document. The requirement that the transfer be in written form does not apply in a case of legal change of copyright ownership or provided that a third party is acquiring only a non-exclusive licence. It is possible, but not an absolute legal requirement, to record the copyright transfer (or any other disposition over it, e.g. assigning non-exclusive licences) with the US Copyright Office (17 U.S.C. § 205). This so-called "recordation" (of transfers or other documents) offers various legal advantages in association with copyright registration.

Apart from the fact that before 1 March 1989 the owner of the copyright was unable to pursue infringements without recording the transfer, recordation gives constructive notice of the facts stated in the recorded document. Recordation means that any claim of ignorance about the ownership of the registered work would not stand up in court. Recordation of the transfer can also establish who should have first priority in using the work if various acts of transfer have occurred and several persons are asserting claims over a work.

Utility contracts / licences

The cession of a design for further use is based upon the granting of a licence and the **corresponding grant of rights** of use. The originator may concede rights of use for his/her work to a third party (the licencee). The concession of rights and the granting of a licence are carried out by means of the so-called "**utility contract**" (licence agreement). Within the legal framework of this contract, the rights of use for the specific rights of exploitation are then conceded and granted.

Tip: *So-called collecting societies represent the interests of their members in respect to all legal rights and claims arising from creative activities, which the originator cannot pursue him/herself for practical or legal reasons. Among the latter would be the collection and distribution of utility fees for the use of works protected by copyright. Moreover, collecting societies – in their function as "lobbying organisations" for originators – typically exert influence on state legislation. The various collecting societies co-operate with one another on an international basis* (→ p. 280, Tips and Links).

Personal Copyright. German originators (→ p. 244, Note from the Legal Translator) are also entitled to a personal copyright, especially the right to the recognition of his or her authorship, and the right to be named as the originator of the work (German Copyright Law, "**Recognition of Authorship**" § 13):

"The originator has the right to recognition for his/her authorship of the work. He/She may determine whether the work is to be given a title of authorship, and which titles should be employed."

Such agreements may be made on a contractual basis or tacitly. The right to name the creative product may indeed be relinquished, but this waiver can also be rescinded at any time. The originator may furthermore determine whether and how the work is to be published (**publication rights** § 12).

In the Federal Republic of Germany the **right to one's own image** is put forth in the **Copyright Law for Works in the Fine Arts and Photography** (German: Kunsturhebergesetz).

Identification systems for digital image files

It has become the standard for international press photographers to identify digital image files with the use of an accompanying text carrying a copyright stamp in IPTC format. The **IPTC** (International Press Telecommunication Council) (→ p. 280, Tips and Links) format is an internationally valid norm with which images are "sloganed" and other important file data can be transmitted, e.g. the copyright notation.

Photo agencies

Photo agencies – also photo archives, picture archives –
market rights of use for images, pictures, photographs, illustrations, as well as film material ("footage"). Basically, photo
agencies offer visual material which is **subject to licensing
regulations** (also "**rights-managed**" or with the acronym
"**RM**"), but which is erroneously referred to as "**royalty-free**"
with "**RF**".

Whereas – in the case of visual material subject to licensing regulations – a licence fee is charged for each usage
according to type, extent and purpose, royalty-free images
are licenced only once, and can then be utilised in a temporally unlimited capacity and in various media. Along with
their generally attractive price models, simple licensing, as
well as the temporally unlimited use of one-time licensed
material, are the most frequently chosen options.

The protection of ideas

In general, ideas as such cannot be protected, for the protection of ideas is neither set down in the copyright laws of individual countries, nor is it provided for in respective state legislation. Merely the external (perceptible and expressed) form,
the design, can be protected, not the idea itself or its content.

The **imitation** of an idea can, however, constitute an
action transgressing competitive rules in a given case
(→ p. 275, Competition Law) and could – according to German
law, for example – violate the Unfair Competition Act § 1
(→ p. 276) (→ p. 244, Note from the Legal Translator). It could
also be the case that – following the termination of unsuc-
cessful contractual negotiations – the individual to whom an
idea or concept has been entrusted during those negotiations
might attempt to misuse the information for personal gain.

During the unique setting of a **competitive presentation**,
the so-called "**pitch**", competition protection could also
become necessary under particular circumstances. It could
be the case that the originator of an idea is forced to demand
forbearance and compensation for damages from imitators or
plagiarists. Practical experience shows, however, that proving
such claims can be somewhat complicated, to say the least.

Tip: One method of obtaining protection from imitators and plagiarists is to insure oneself against the unauthorised exploitation or cession of the idea, or to guarantee the confidential treatment of such information entrusted to others, by including these points in the negotiated contract. For this purpose, a **non-disclosure and confidentiality agreement** *(NDA) should be drawn up with the potential client, including the condition that any breach of the agreement be punishable by law, with the understanding that a failure to sign the agreement entails the refusal of the commissioned party to reveal the substance of an idea or the precise contents of a conception. If it should nonetheless come to a violation of the contractual stipulations negotiated beforehand, the party suffering damages has the right to demand compensation as a consequence. Similarly, the conceptual development of an idea may be rendered in written form, as well as certified by a notary public or by written proof of committal.*

Protection Rights for Industrial Property

Commercial rights of protection coincide with so-called "industrial property". This concept encompasses the legal coverage of intellectual accomplishments in the commercial sector such as patents, industrial designs, brands, pattern designs, etc.

Registered design law and design patent law

Registered design law is often referred to as the "little brother" of copyright law. Protection is granted to the external form, i.e. the design. The sole prerequisite is that the creative production be "new" and "unique". Registered design law is a so-called untested copyright, which means the official evaluation is restricted to formal requirements. This has the consequence that registered design law does not presuppose a "formative level".

The legal integrity of a registered pattern and its potential scope as regards the need for legal protection must often be clarified in court for the very first time – within the context of a legal dispute. Nevertheless, it can be worthwhile to register a pattern design, if the design has been fully developed.

Registered design in the Federal Republic of Germany offers holders a high degree of protection with a relatively low amount of financial and temporal expenditure, as the period of legal protection can range up to a decade (10 years). An official entry can be obtained rather quickly, as a rule, within three to six months following registration, after which it assumes legal validity.

Registered design in Europe. Since 12 December 2001, a European Community directive regulating design has been in force. Its jurisdiction extends to all domestic markets of the member states. The official entry of a design can be accomplished by means of a single registration guaranteeing uniform protection for design production throughout the entire European Union. Further information can be requested from the **Office for Harmonization in the Internal Market** (OHIM). The OHIM (→ p. 279) is the official organ of the

European Union and is responsible for the registration of designs and brands for all member states of the European Union.

Protection of Printed Characters. For all typographic characters there exists a special legal framework represented by the **Typeface Protection Act**. The registration procedure is modelled on the registered design law. A **font or script name** may be protected as a **brand (trademark)**. As long as no trademark protection exists, the use of the original name may – under certain circumstances – constitute an unfair competitive act (→ p. 276). More specifically, this might give the false impression that the source of the script is being treated as the original, which would be tantamount to an unlawful act of deception in respect to the origins of the work. According to the Typeface Protection Act, each individual character cannot yet be considered a pattern, but only the entirety of characters or letters constituting a typeface.

European registered design law corresponds with the copyright protection on **industrial design** in the USA (17 U.S.C. § 1301). This lays down that original design (not just trivial or copied from another source) for a useful article can be granted copyright protection provided that the design makes the product attractive or distinctive in appearance to the purchasing or using public. Conventional or merely functional designs are not subject to protection on these terms. The term of protection continues for 10 years after the design is first made public or after published registration, whichever is the earlier. Unlike copyright, industrial design must display a "design notice" similar to a copyright notice (the words "Protected Design", the abbreviation "Prot'd Des", the letter "D" in a circle or the symbol "->D->", together with the year in which protection commenced and the name of the owner of the industrial design), and be registered two years after it is first made public, in order to be able to claim rights in the design. Industrial design is the simplest and commonest way of protecting designs in the USA. It was introduced to promote creative, decorative production and to enhance and at the same time protect the beauty and saleability of industrial products.

A design can also be protected in the USA under patent law by a so-called **design patent** (35 U.S.C. § 171). But this does not have the same significance in practice, as it is considerably more difficult, more expensive and takes longer to acquire a design patent. As a rule, patenting industrial designs fails because they lack patentability. The threshold for patent protection is very high, as the legal patent requirements of novelty, usefulness and non-obviousness have to be met, and this very rarely happens. Patenting a design is also very expensive, as it is impossible to register without patent lawyers.

Trademark rights

A **trademark** facilitates the identification and public recognition of goods and services originating from a particular business enterprise. Thus, it likewise facilitates the distinction between the various commercial origins and diverse providers of goods and services. Frequently, brands are indicated with a "®" or "™" (according to US law, as these are irrelevant in Europe).

The trademark sign "**TM**" (fig. 4) is a legal designation used predominantly in the USA. The mere use of the TM abbreviation, however, does not constitute the actual legal claim. The abbreviation is only a visible indication that viable legal claims in relation to this particular trademark exist. It is not unusual in the USA for many brands to carry the TM abbreviation for which trademark protection has been applied for, but not yet granted.

"**R**" = Registered (fig. 5) designates brands which are both officially registered and legally protected. The trademark holder has thus acquired the liberty to situate the ® in any desired position. Nevertheless, the rights stem solely from the registration entry and the registration document.

The most frequent types of **brands** are **word marks** (brand names), **design marks** (brand figures/logos) or hybrid forms such as **word and design marks**. Beyond these, the past years have witnessed the development of further brand types such as **colour marks**, **olfactory marks** and **sound marks**.

Trademark protection is established after a completed registration procedure involving the **entry of a trademark** into the respective goods and services categories.

4

TM

5

®

Additionally, in the USA, the Federal Republic of Germany and other countries, there exists trademark protection without any prior registration entry, e.g. by dint of public recognition, or through the mere use of the brand.

Tip: *A trademark should always enjoy formal legal protection.*

Tip: *To the extent a trademark design is being created, worked on and/or commissioned, the designer should negotiate a written agreement with the client or commissioning party to the effect that no guarantees will be given regarding the eventual registration of the trademark. The following text can be adapted for this purpose: "The designer is not liable for either the reliability of the commissioned work in respect to competition law and trademark law, or its suitability for registration purposes."*

How does a trademark become registered?

Principally, the designer has the following **trademark protection system** at his/her disposal:

1. The registration of a **national trademark** (country).

   The period of protection for a registered **German trademark** ("DE-Marke") begins on the day of the filing of the application and terminates 10 years following the end of the same month. The period of protection may be extended by another ten years, respectively. According to German trademark law, the following trademark types may be placed under legal protection: brands and trademarks, business designations (corporate trademarks/work titles) as well as geographical source information.
   Unlike copyright or patent law, **American trademark law** is not subject to federal legislation alone. So alongside the federal "Lanham Act", each American state has its own trademark law ("state common law"). Thus a distinction is made between "common law trademarks" – trademarks that are used commercially without being registered with the United States Patent and Trademark Office (→ p. 279, Tips and Links, US PTO) –

and "registered trademarks" – trademarks that are registered with the US PTO. Registering a trademark is recommended as it affords the owner of the trademark more protection and offers numerous advantages in terms of procedural law.

Basically, the following may be registered as a brand or trademark for any services and commodities whatsoever: words, logos, three-dimensional figures, jingles (i.e. brief, memorable tone sequences or melodies), colours, as well as combinations of all the foregoing.

Trademark protection is to be applied for at the respective Patent Offices (→ p. 279, Tips and Links).

2. The registration of a **European Union trademark (EU trademark)**

The "Community" or European Union trademark offers the corresponding legal protection for all **member states of the European Union**. Basically, words, logos, three-dimensional figures, colours as well as combinations of all the foregoing may be registered as a brand or trademark. Given that no formal obstacles exist, the application will be published approximately within the following nine months after the filing of the application. Afterwards, it is possible for the holders of pre-existing trademark rights within the European Union to raise any objections. Should no objections be raised, the official registration is then carried out. Here, the period of protection covers ten years and can be extended on request for another ten years, respectively. The Office for the Harmonization in the Internal Market (OHIM), located at the seaport Alicante, in southeast Spain, is responsible for the evaluation of EU trademark registrations (→ p. 270, Registered design in Europe).

Tip: *Please be aware of the fact that – in contrast to an international, i.e. "IR" trademark registration (→ p. 275) – a single objection can endanger the entire process of EU trademark registration. Should the legal objection be successful, the EU trademark may still be reworked for other national markets, respectively. The disadvantage: The costs can be relatively high.*

3. The registration of an **international trademark
    (IR trademark)**

    The international registration of a trademark is regu-
    lated in accordance with the **Madrid Trademark Agree-
    ment** (MTA) as well as the Protocol of the Madrid
    Trademark Agreement (Madrid Protocol), and provides
    for central registration through the World Intellectual
    Property Organisation (WIPO) in Geneva (→ p. 279),
    servicing different countries simultaneously. Along with
    their European partners, member states of the Madrid
    Trademark Agreement and the Madrid Protocol (approx-
    imately 78 countries to date) also include Japan,
    Australia and the USA. The period of protection covers
    ten years and can be extended on request for another
    ten years, respectively. The prerequisite for international
    registration is the simultaneous existence of an identical
    trademark that has been registered in the home country
    of the petitioning party ("home registration").
        In principle, slogans and claims – so-called combined
    word marks – may also be entitled to trademark protec-
    tion and are subject to the selfsame evaluative criteria.

Protection of titles

Trademark law provides separate protection in the form of
work title protection, or title protection for names or special
designations of printed texts, auditory works, stage productions and other comparable works.

Competition law

In response to violations regulated by unfair business practice laws, the injured party may demand forbearance and
compensation for damages.
    In all **member states of the European Union** – with the
exception of Great Britain and Ireland, which do not have
unfair business practice laws in their respective legal systems – the legal protection of designs against unfair commercial competition is enforced at the national level. Currently,

at the European level, there exist no legal provisions to regulate unfair business practices. In order to achieve partial harmonization, the Council of the European Community has indeed issued normative directives (84/450/EEC from 10 September 1984 as well as RL97/55/EC), but wherever legal jurisdiction has not been assumed by EC law, national laws remain decisive.

Tip: *The designer is not indeed fundamentally liable for either the admissibility of the commissioned work in respect to competition law and trademark law, or its suitability for registration purposes. It is nevertheless advisable to agree on a legal proviso with the commissioning party as follows (→ p. 273): "The designer is neither liable for the admissibility of the commissioned work in respect to competition law and trademark law, nor its suitability for registration purposes."*

Comparative advertising is advertisement which either directly or indirectly exposes a competitor, as well as its services and commodities. Provided that comparative advertising is not false or misleading it is largely admissible in the USA, even if a competitor's product is ridiculed.

In the Federal Republic of Germany (→ p. 244, Note from Legal the Translator), comparative advertising was – with few exceptions such as test comparisons – principally forbidden. As of 1 September 2000, subsequent to amendments to the German Unfair Competition Act (UCA) and as a response to the **EC Comparative Advertising Directive**, such advertising has indeed become basically permissible, but only provided that those prerequisites put forth in § 6 **Unfair Competition Act** are adhered to. Comparative advertising is basically permissible if it is not immoral or if it does not offend public sensibilities. What may be considered offensive in this context can be found in the Censorship Catalogue provided in UCA § 6, Para. 2.

Caution is advisable in connection with **price comparisons** used for the advertisement of special offers: Here, German law (UCA § 6, Para. 3) requires that the period of availability for the special offer be indicated clearly and unmistakably.

Internet advertising

The **member states of the European Union** have reached a consensus within the legal framework of their **Directive on Electronic Commerce** (2000/31 EC) that – as regards the electronic commerce (**e-commerce**) sector – the legal principle of national ascendancy, i.e. the priority of national law, should prevail. Correspondingly, any offerer of a **website** must observe certain legal obligations.

Offerer identification. Every person/entity presenting itself/themselves on the internet must publish information concerning its own identity, e.g. in the **Federal Republic of Germany** (→ p. 244, Note from the Legal Translator), since 21 December 2001 in accordance with the Electronic Commerce Law, with additional duties of identification in accordance with the Teleservices Act § 6 (German: Teledienstgesetz).

Tip: *Internet designers and computer programmers alike should follow these developments closely; legal disputes involving authorship, trademark or lawful competition can prove to be very costly for the potential client or commissioning party, and should be avoided for these reasons as far as possible.*

In the Federal Republic of Germany, the so-called **imprint obligation** (→ p. 244, Note from the Legal Translator) lays down the legal obligations of proper identification for offerers a website. Every party offering services on the internet is obligated to present specific information such as a full address, telephone number(s) and/or fax number, as well as the value-added tax identification number.

Naming a post box alone is insufficient, just as merely naming an e-mail address is inadequate. Furthermore, the names of natural and legal persons, or business partnerships with corresponding representational authorisation and duties, e.g. Inc., Ltd., etc, must also be indicated. When indicating the address, it is important to include all necessary functional data such as postcode, location, street and house number.

Tip: *Admittedly, while an internet designer is not liable for false or incomplete information in the commissioned work, he/she should nonetheless inform the client about all necessary legal formalities.*

Domain titles/names. Name holders or companies have a fundamental right to a corresponding internet address. This means: As a rule, these parties are entitled to exclusive rights in connection with the use of their own business name or product trademarks. Anyone planning to reserve a domain title should, however, do thorough research. Otherwise, the commissioning party – or even the designer him/herself – could become involved in costly forbearance proceedings with the holders of the contested trademark or name rights.

    Links and frames. In order to avoid the hazards connected with authorship and liability rights, the use of links should be well considered. Every complex website is protected by copyright. The average website link is usually unproblematic in respect to copyright considerations. Whoever publishes work on the internet simultaneously expresses tacit approval that cross-references to that work will be made on the part of other internet users. As far as tacit approval is concerned, so-called **deep links** should be carefully examined, which refer to an underlying group or entire complex of other websites.

    Inline links – whereby the linked website appears within the **frameset of the main provider** – are not permissible without the express consent of the originator. This is due to the fact the "average" internet user cannot recognise that the contents actually belong to another internet presence.

Tips and Links

Law

<u>USA</u>
Office of the Law Revision Counsel of the US House of Representatives (only U.S.C., i.e. the codification of US Federal Law without decisions by the Federal Courts and without the law of individual US states): uscode.house.gov

National Archives and Records Administration (Code of Federal Regulations, i.e. rulings by federal administrative authorities): www.access.gpo.gov/nara/cfr/cfr-table-search.html

<u>Europe</u>
Access to European Union Law: www.eur-lex.europa.eu

<u>Germany</u>
Federal Ministry of Justice/Laws (in German): www.bundesrecht.juris.de

National and International Offices and Authorities

<u>International</u>
World Intellectual Property Organisation (WIPO): www.wipo.int

Universal Copyright Convention (UCC): http://portal.unesco.org/culture

<u>USA</u>
US Patent and Trademark Office (US PTO): www.uspto.gov

The United States Copyright Office: www.copyright.gov

<u>Europe</u>
European Patent Office: www.european-patent-office.org

Office for Harmonization in the Internal Market (OHIM): www.oami.europa.eu

European Commission: www.ec.europa.eu/internal_market www.ipr-helpdesk.org

<u>Great Britain</u>
Trade Mark Text Enquiry UK Patent Office (UKPO): www.patent.gov.uk

<u>Germany</u>
German Patent and Trade Mark Office (in German): www.dpma.de

National and International Associations

<u>International</u>
Art Directors Club (national chapters exist in many countries worldwide): www.adcglobal.org

International Council of Graphic Desgin Associations (ICOGRADA): www.icograda.org

<u>USA</u>
American Institute of Graphic Arts (AIGA): www.aiga.org

<u>Great Britain</u>
British Design Council:
www.designcouncil.org.uk

Collecting Societies

<u>International</u>
International Confederation of Societies of
Authors and Composers (CISAC):
www.cisac.org

<u>USA</u>
Artists Rights Society (ARS):
www.arsny.com

<u>Japan</u>
Japan Artists Association Inc. (APG-JAA):
www.jaa-iaa.or.jp

<u>Great Britain</u>
Design and Artists Copyright Society
Limited (DACS): www.dacs.org.uk

<u>Germany</u>
Verwertungsgesellschaft Bild-Kunst (VG
Bild-Kunst): www.bildkunst.de

Other

International Press Telecommunication
Council (IPTC): www.iptc.org

The Pixelboxx IPTC Writer for the digital
titling of photo files (freeware or full version): pixelboxx.com/iptcwriter/index.jsp

9.0

As far back as the 17th century, Shakespeare detected "method" in Hamlet's "madness". This astute – and timeless – observation still holds true: a modicum of organisation can add the necessary order and purpose to our creative chaos and, when it comes down to it, help us attain our goals.

In an age when most people waste around 80 per cent of their time on meaningless tasks, those who know how to use their skills, experience and tools to their best advantage work far more effectively and can even improve the intricate choreography of a large-scale project. With minimal planning and – sorry, guys – a healthy dose of self-discipline, imposing a basic system and order on one's own routine becomes child's play. And this is a wise and even lucrative move because even the most wonderful design is bound to fail if it goes over budget, misses its deadline or fails to meet the client's stipulated requirements.

So, stop dithering and get right down to the nitty gritty – turn theory into glowing design practice with our structural aids to face and outfox those pesky demons of chaos and procrastination. In addition to guides for organising and planning your projects and work, this chapter also packs in plenty of tips and tricks on time management, presentation techniques and honing your rhetorical skills.

And don't worry that you might spend too much time on planning. As soon as you add a little structure to your day, the improved organisational framework will actually free up more time and leisure for ambitious creative explorations, indulgences and flights of fancy.

Organisation

9.1	**Planning Jobs and Projects**	284
	Planning jobs	284
9.2	**Project Planning**	287
	The project plan	290
	Communication in project management	294
9.3	**Sample Plans**	296
9.4	**Presentation and Rhetoric**	297
	Checklist: Creating a presentation	297
	Rhetoric	298
	Planning a presentation	300
	Designing a presentation	301
9.5	**Tips and Links**	305

Planning Jobs and Projects

Planning is a set of systems and methods. Good planning is purposeful and clear-sighted, effective and efficient; it helps to avoid mistakes.

<u>Key planning questions</u>

- What am I trying to do? (Aim)
- What is important? (Criteria)
- How do I best go about achieving the aim within the given conditions? (Working method)
- When do I start? (Deadlines and time)

Golden rule: Planning should always be done in writing. Too little planning creates problems; so does too much planning.

<u>Aims</u>

Aims are a decision-maker's guidelines and signposts. You can't set priorities without defining an aim. A well-defined aim is realistic and concrete. It is not too ambitious, not too modest, and sets a deadline. Aims should be verifiable, realistic, consistent, complete, achievable and transparent.

Golden rule: Aims should always be put down in writing. An aim that has been written down is a commitment, to yourself as well, to address a task with determination and vigour. Aims should be checked regularly and amended if necessary.

Planning jobs

Every project needs a plan. All the jobs to be done should be listed in writing, and the list kept at hand. Jobs can be listed on paper or digitally, using a computer or a PDA (Personal Digital Assistant), as preferred.

The first aim is not to do all the jobs that have been noted down immediately, but to place them in rank order. The important jobs should always be done first. And making the list is not enough, however good it is. You have to work with it.

Tip: *Jobs that have not been noted down or included in the plan chase each other around in your mind and get you down. If you write something down on your list of jobs, that's a commitment: these jobs have to be done! Write down the next concrete step for each job, so that you have a clear sense of what's to be done, and what's to be done next. And at the same time remember that there are always more jobs than time available.*

<u>Golden rules of job planning</u>

- Self-discipline.
- Little jobs needing less than three minutes are always done at once.
- Larger jobs should be broken down into several steps.
- Check the job list once a week.
- Set up a calendar including times for follow-up.
- Keep a deadline reminder (e.g. calendar, reminder function on MS Outlook, iCal, resubmission folder).

Note: Unless you have some kind of reminder system you will be stressed and tense because you constantly feel you must have forgotten something.

Tip: *Plan the evening before and let your subconscious work for you.*

Self-discipline means being able to overcome our innate laziness and inertia, and to do even unpleasant tasks in order to achieve more in the end.

Tip: *Constantly ask yourself: "What will I gain by doing that now?"*

Setting priorities. Usually we squander 80 per cent of our time on work that gets us next to nowhere. The 80:20 principle states that for most people only **20 per cent** of their work leads to about **80 per cent** of their success and productivity.

Tip: *Identify the most important jobs on your job list. Avoid disturbances and interruptions. Set priorities: "very important" (A), "important" (B) and "routine, can be done later" (C).*

Time management

There is a lot of sense in the saying "Time is money". Time has to be organised so that it can be used well. An imprecise estimate is better than no estimate at all.

Every job, activity and deadline should be written down at once. That's the only way to make sure you're on top of every situation. And it's the only way to concentrate on essentials.

Tip: *Don't plan the whole day away. Allow a buffer zone of about 40 per cent. Plan three blocks of time: about 60 per cent for planned activities, about 20 per cent for unexpected activities ('thieves of time') and 20 per cent for spontaneous and social activities. Experience will show what can be done and planned in a day, and what is too ambitious.*

Note: Your inner attitude affects everyone around you, not just you. Think positively!

Project Planning

A project is a **job** or a **plan** that has a **defined beginning** and a **defined end**. It requires several linked **sub-projects** that have to be gone through in order to achieve the **given aim**.

Every project is based on a concept. The **concept** is a comprehensive intellectual design based on a key idea or particular aims. It brings fundamental action and strategy frameworks together, and also the necessary operational actions.[1]

Any project requires **organisation**: creating and carrying out a plan.

<u>Four golden rules of planning</u>

1. Analysis: What is the problem or what needs to be done?
2. Planning: How do I tackle the problem or what needs to be done?
3. Implementation: What does the solution look like?
4. Monitoring: How do I check the outcome?

Projects usually fail because people rush into the work without planning in advance, because deadlines are missed or postponed, or because individual jobs were not adequately monitored.

Projects usually succeed if:

- the project leader and the team have the same clearly defined aims and outcome in mind;
- the project is adequately planned, above all to prevent false starts and having to repeat individual steps;
- the work is carefully timetabled and monitored to ensure the project can be concluded;
- the team members understand what is expected of them;
- there are open channels of communication at all levels and at all business locations;
- the team members are empowered to act and motivated to strive for a high level of achievement, and to sustain this;
- the team members are given emergency plans they can fall back on if events do not run as planned.

---

[1] Becker, Jochen: Marketing-Konzeption. Grundlagen des strategischen Marketing-Managements (1993). No English version available.

Eight golden rules of the planning process

Tip: *Visualise!*

1. Plann in adavance.
2. Set aims.
3. Think through tactical and strategic alternatives.
4. Fix key elements for individual phases.
5. Plan a timetable.
6. Decide what resources are needed.
7. Draw up a budget.
8. Fix the key principles and procedures.

Checklist: What must the planner and procedures be able to deliver?

- **Financial viability** for sub-projects and the project as a whole.
- **Reliable planning** (of staffing, costs etc.).
- **Clear presentation** of the workflow with all the key dependent factors.
- **A detailed time plan** setting so-called milestones, for example, critical situations and spare time (buffer zones).
- Reliable and simple **documentation** for preparing the work.
- The ability to identify **deviations** from the plan rapidly and respond to them in good time.
- Integration for all **phases** of the developing project (design, production, purchasing, finance etc.).

Checklist: The planning process

Planning

1. Draw up an initial **project plan** and publish it.
2. Define aims.
3. Structure jobs (sub-jobs).
4. Determine connections.
5. Plan jobs.
6. Plan resources.
7. Draw up plan.
8. Release plan for action.

Tip: *A project plan should be devised for every project and defined in writing. The first plan fixes the budget and the duration of the project and identifies responsibilities. This plan contains the following: project description, project duration, project budget, project team and project outcome.*

Implementation

1. Check aims.
2. Allocate jobs.
3. Motivate.
4. Do the work.

Monitoring

1. Monitor work activity and workflow.
2. Follow progress.
3. Keep an eye on tendencies and events.
4. Report outcome.

Adaptation

1. Solve problems.
2. Modify the plan if need be.
3. Fall back on emergency strategies.
4. Conclude project.

The project plan

How is a project structured?

Draw up a project structure plan. This **structures activities** and other project elements clearly before the actual sequence of events is worked out with the aid of an **activity plan**, the **timetable** and **cost plan**. The project structure plan places the individual work bundles in a hierarchical structure and gives a complete overview of the project. Then the work bundles are broken down into individual procedures and built into the project workflow.

Note: Plan before the project begins, and again if new aims are introduced. Adapt the plan if results are unsatisfactory and if requirements, resources or deadlines change.

The following summary (fig. 1) shows a function-driven project structure plan.

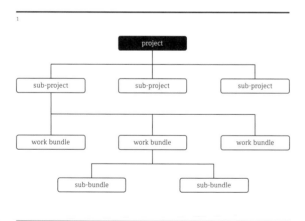

The project plan is a detailed description of the project design. A project plan is made up of the following parts:

1. **Project definition**
   - Description of the work or jobs to be done
   - Project requirements
   - Project aims and project commission
   - Project outcome
   - Expectations (final outcome and aims)
   - Scale of work (internal and external)

2. **Project variables**
   - Work to be done
   - Project starting date
   - Predicted project duration
   - Planned conclusion date
   - Expectations (final outcome and aims)
   - Personnel requirements
   - Implications

3. **List of milestones and work to be done**
   - Responsibilities
   - Task
   - Expected outcome
   - Planned starting date
   - Planned finishing date
   - Actual starting date
   - Actual finishing date

4. **Project budget**
   - Actual budget
   - Projected budget
   - Further budget expenses (e.g. travel expenses)
   - Unbudgeted costs

5. **Supplementary plans**
   - Training
   - Implementation
   - Emergency plans

6. **Project approval**

<u>Types of plan</u>

A **flow plan** illustrates the logical connections between the defined work bundles (jobs). The network planning technique supports this illustration with targeted **graphs**; here the jobs are presented as nodes in a network or as logically interdependent connections (e.g. related to time).

The **timetable** supports project timetabling by determining how long each job will last, the order in which jobs are done and resource allocation. Other types of plan are the **capacity plan**, **cost plan**, **staffing plan**, **quality plan**, **financial plan** and **risk analysis**.

<u>Striking visual presentation for your activities</u>

Large to middle-sized projects are presented visually using **network plans** (network planning technique based on events and activities). These show project timing graphically. Project sections at a particular point in time are identified by circles. Activities are plotted in the form of lines ending in arrows.

The following graphic (fig. 2) shows the **PERT method** (Programme Evaluation and Review Technique).

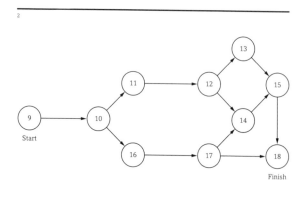

Middle-sized to **small projects** can be presented as follows using the so-called **Gantt method**,[2] (fig. 3) structured on the basis of Who? When? What? and How? Lines in a matrix are set against a time axis.

3

Plan of work								
Job	Responsible	Week (Calendar Week)						
		2	3	4	5	6	7	8
01 Draw up project plan	JM	▬	▬					
02 Seek approval	MM		▬					
03 Work out strategic plan	SB		▬	▬				
04 Install hardware	Supplier			▬	▬			
05 Book project staff	Admin		▬	▬	▬			
06 Write training module	JB			▬	▬			
07 Test system	MM				▬	▬		
08 Project staff start work	Admin						▬	
09 System launch	RK						▬	
10 Training	JB						▬	▬
11 System switch	CM							▬

Smaller projects should contain a list of milestones and work to be done as follows:

– Responsibilities
– Job
– Expected outcome
– Planned starting date
– Planned completion date
– Actual starting date
– Actual completion date

Developed by Henry Laurence Gantt around 1910.

Checklist: Project planning phases

1. Make preliminary decisions
2. Analysis
- Project analysis, time analysis, necessary cost, materials and capacities analysis
3. Structure planning
- Systematic presentation of the project structure
4. Flow plan
- Systematic presentation of the project flow
5. Time planning
- Concrete preliminary planning for the timetable
6. Revision
- Required if adjustments are needed

Communication in project management

Efficient communication is essential for project management. Efficient communication means, above all, ease of understanding. This can be described in terms of the following four characteristics using the Hamburg Model (fig. 4) devised by the German communications specialist Friedemann Schulz von Thun:[3]

- Simplicity not complexity
- Structure and order not chaos
- Succinctness not waffling
- Stimulating additions not mere information

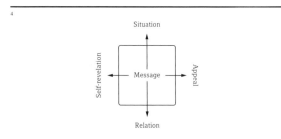

A successful communicator is aware that a message can communicate several things on different levels (→ ch. Design, p. 34) and can also be conveyed via several communication channels.

Checklist: Project monitoring

1. Keep an eye on milestones.
2. Hold regular team meetings.
3. Include time buffers in planning.
4. Allow for unforeseen events in all important activities.
5. Have a carefully structured emergency plan at hand.
6. Do not tolerate inactivity.
7. Keep a project diary.

Advantages: A project diary records things that have been agreed so that important project information is always available. It stops you from remembering only positive situations. It is also useful as a sequential description of the work, and for writing the final report.

Note: People feel more committed to something they have put in writing than something they simply remember.

8. Write interim reports
   – containing the date of the report, the project starting date and the probable completion date;
   – recording the flow of events, the time taken and revised calculations;
   – Listing points (jobs) that are to be decided;
   – containing recommendations about next stages or termination of the project;
   – fixing the date of the next report;
   – containing additional important points and the author's contact data.

The **final project report** presents a final retrospective assessment listing mistakes, results, successes, improvements, time schemes for future projects and suggestions for improving project management methods.

Sample Plans

Scheduling Template

Client:		Project Manager:	
Project Number:			
Project Name:			
Date:			
Project Team:		Internal:	
		External:	

Project Phases with Description:	Date		Responsibilities
	Start	Finish	
1. Start of Project and Definition of Tasks			
2. Concept Development and Tuning Phase			
3. Design and Realization Phase			
4. Implementation Phase			

Annual Plan		2007	January					February				March				April				May					June				July				August				September				October				November				December			
	CW		1	2	3	4	5	6	7	8	9	10	11	12	13	14	15	16	17	18	19	20	21	22	23	24	25	26	27	28	29	30	31	32	33	34	35	36	37	38	39	40	41	42	43	44	45	46	47	48	49 50 51 52
Instruments/Phases	Budgeted Costs in €																																																			
1	0.00																																																			
2	0.00																																																			
3	0.00																																																			
4	0.00																																																			
5	0.00																																																			
6	0.00																																																			
7	0.00																																																			
8	0.00																																																			
9	0.00																																																			
10	0.00																																																			
1	0.00																																																			
2	0.00																																																			

Presentation and Rhetoric

Presentation is a way of using purposeful **recipient-oriented design of data**, **facts** and **statements** to make information more efficient.

Any good presentation is directed at the **recipient**, so the following should be checked before every presentation:

Checklist: Creating a presentation

– Who are the addressees, also called target groups or participants? Who is the presentation intended to reach? What prior knowledge about the subject can be assumed? What is the approach to the subject? What will be the likely readiness to work on it, general interest and number of participants?
– What is the presentation intended to achieve?
– What are participants intended to know or do when the presentation is over?
– How is the presentation expected to change the participants?

The presentation should contain the following:

– **Key statements** that cannot be omitted.
– **Important statements** to flesh the topic out.
– **Interesting statements** to add flavour to the topic.
– **Examples** (e.g. case studies or experiences) illustrating the topic.

Tip: *The presenter should always have sound* **background information** *at hand, covering the topic fully and providing answers to any possible questions.*

Rhetoric

Rhetoric is the **art of speaking** and, at the same time, of **analysing language**. It also deals with effective communication and using convincing arguments to persuade listeners and spectators to change their minds, or to act in a certain way.

Rhetoric originated in the **ancient world**, in the fifth century BC. It was not usual to be represented by a lawyer in court at this time. Both plaintiff and accused spoke on their own behalf, and the case was awarded to the person who spoke more convincingly. The Greek philosopher and naturalist Aristotle (384–322 BC) first presented a systematic analysis of the art of speaking. "Rhetoric" is one of his major works.

Renaissance scholars were familiar with Latin speeches. The Humanists modelled themselves on the rhetoric of the Roman statesman and philosopher Cicero (106–43 BC) and the Roman rhetorician Quintilian (c. AD 35 to c. 96). Modern political rhetoric evolved above all in the context of British parliamentarianism and through exponents such as William Pitt, Edmund Burke, Charles James Fox, Richard Brinsley Sheridan and Winston Churchill.

Communicative elements in a presentation

- **Personality**, through appearance and manner, for example.
- **Body language** such as facial expressions and gesture.
- **Words**, spoken, written or through the choice of vocabulary.
- **Presentation** style, e.g. visual aspects and sound.

Golden rules of rhetoric

1. Speak factually only where essential.
2. Speak coherently.
3. Speak distinctly, and not too quickly.
4. Pause (to give your audience time to absorb the material).
5. Do not read charts or projected material aloud.
6. Use cultivated language.
7. Speak as freely as you can; do not simply read a prepared text.

8. Use the so-called crib-sheet method, which will allow you to speak as freely and confidently as possible.
9. Prepare your arguments and material thoroughly.
10. Argue convincingly, objectively and fairly.
11. Speak with a clear end in mind, and structure your material lucidly.
12. Simplify complicated material.
13. Come to the point.
14. Check foreign words, technical terms and numerical material,
15. Maintain eye contact and take listeners' feedback into account.
16. Remember body language.
17. Keep your facial expressions and gestures natural.
18. Make a reliable impression with your speech and manner.
19. Open up to your listeners, move towards them; adopt a reasonably familiar tone.
20. Listen actively and analytically, and respond to questions.
21. Address the listeners' requirements and expectations.
22. Avoid making undue demands on your listeners, or overtaxing them.
23. Don't speak for longer than your listeners' attention span allows.
24. Capture their interest and introduce the subject gradually.

<u>How do I make contact?</u>

– Address participants or people you are talking to directly.
– Maintain eye contact.
– Look at individual participants directly.
– Open gestures: open your arms and hands to the audience, don't fold them.
– Keep your movements economical and calm: change your posture or approach the group.
– Attract attention with variations in your voice: articulation, pitch, pace, volume and tone.
– Choose appropriate media to illustrate your subject: vivid language, use of examples, demonstration objects etc.

Body language

Body language can be changed only to a limited extent.
It is expressed through posture, eye contact, and vocal traits
like volume, tone, modulation, speech tempo, pauses, distance (angle) between you and the people you are talking to.

Tip: *The first impression is crucial, the last impression stays
with the audience.*

Planning a presentation

Presentations are planned with a script in the form of various
schemes such as the linear method, a tree structure or a net
structure (→ ch. Digital Media, p. 144, 145, figs. 11, 12).

Checklist: Planning schedule for a presentation

1. Fix the presentation aims.
2. Fix the presentation strategy, bearing the target group
   in mind.
3. Fix media and layout.
4. Structure the content and put background material
   in place.
5. Create script.
6. Devise texts and images.
7. Production.
8. Trial run with people under realistic conditions.
9. Presentation.
10. Retrospective analysis and evaluation.

Crib methods

Bullet point manuscript. Bullet points provide continuous
hints and information that are then put in your own words,
explained or expanded with examples.
    Advantages: Speaking freely, direct contact, e.g. eye
contact, with the audience. This makes the presentation
seem lively and natural.

Tip: You should write out the first sentences of the introduction, the summary and the conclusion in advance. You can then be sure of finding the right words at nerve-wracking moments at the beginning or end of a presentation, for example.

File cards. It is best to use file cards not larger than DIN A5 (→ ch. Production, pp. 153, 154). File cards are stiff and easy to handle, do not rustle, obstruct very little of the speaker from view, and do not distract either spectators or listeners. They also keep notes down to the most important points.

Tip: Use different colours, e.g. white cards for essential material and yellow cards for material that can be omitted if time is too short, for example.

Designing a presentation

The right **use of media** or **presentation tools** can strengthen a presentation (→ ch. Digital Media, p. 111, Hardware and Software).

The **design resources** available for a presentation include typeface, type size, line length, line spacing, surface, colour, contrast, order, animation, sound. Remember that a first, cursory glance can determine whether a presentation will succeed or fail.

Tip: Important material should be placed in the primary and secondary levels of information.

Digital presentation charts

The most frequent mistakes made in designing and using presentation charts are listed below:

– The charts are hard to read; the font size is too small, the contrast too low.
– Individual charts contain too much irrelevant information; the presentation is cramped.
– The design resources do not illustrate the theme clearly.

– The projected material is poorly explained or there is insufficient time available.
– The presentation includes too many or too few charts.
– The charts do not fit in with the organisational scheme of the talk; they do not match the particular situation; lack of co-ordination between presentation and talk.

Here are some helpful suggestions for designing and using charts:

– All the charts should use the same design grid.

Tip: *The charts should be numbered in sequence.*

– The basic lettering should be at least six millimetres high or the typeface at least 20 point (pt).
– There should be maximum contrast.

Tip: *Lines should be at least 1.5 pt high.*

– The layout should be designed simply and lucidly.
– Less is more! Avoid irrelevant information.

Tip: *Five key statements or five structural points per chart are sufficient. Use keywords, rather than full sentences.*

– Charts should be designed pictorially.
  Note: Avoid too many or unnecessary effects or animations. They make the presentation unsettled and are also greedy of computer memory.
– Colour should be used with a precise intention.
– Content should be easily understood.
– There should be an adequate number of slides to be projected.

Tip: *The maximum number of charts can be worked out by dividing the length of the talk in minutes by three. As a rule, 7 ± 2 charts are used. Three minutes speaking time should be allowed for each chart.*

- The text of the talk and the projected presentation should match.
- Each chart should fit the particular situation, and they should form a coherent whole (uniformity).
- Charts should be shown for sufficiently long, and explained in full.

Tip: *Allow the listeners and viewers time to get used to each chart; announce the chart in advance. Pause for two or three seconds after starting to show the chart; let the chart itself "speak" to the audience first.*

Presentation methods

Complex connections can be shown in various ways in a presentation. Schematic diagrams can take a whole variety of forms for presenting different material, such as figures and the connections between them, in a readily comprehensible way.

For example, the **organigram** presents a set of facts and the way they relate to a complex theme. Connections are illustrated by arrows and branches.

Bar and **pie charts** (figs. 5, 6) are used for comparative numbers. A **schedule** in bar chart form shows start and end dates and presents each activity as bars on a time axis (→ p. 293, fig. 3).

A **pie chart** shows how a part relates to the whole as a percentage (fig. 6). The following table shows various presentation modes and their advantages.

	Simple flow	Comparisons and contrasts	Developments	Structure, composition	Organisational structures	Absolute values	Data allocation	Lists
Presentation type								
Liste						X		X
Table							X	
Bar chart		X				X		
Column chart (→ fig. 5)		X				X		
Pie chart (→ fig. 6)								
Graph (→ fig. 7)	X		X					
Organigram				X	X			
Structure chart				X	X			
Flow chart	X		X					

Checklist: Presentation

1. **Preliminary planning:** How much presentation time and how much discussion time has been allocated?
2. **Check the venue:** Size, acoustics, seating arrangement (Where does the speaker stand?), lighting (blinds, light switch), podium.
3. **Technology:** Overhead projector, beamer, switches, cables, focus, computer, software (sound), laser pointer, flipchart.
4. **Own documents:** Bullet point script, digital presentation.
5. **Participants' documents:** What documents are to be given out to the listeners or spectators?

Tip: *Participants' documents, often called* **handouts**, *should not be distributed until after the presentation. No one likes speaking to an audience rustling their way through press packs or other documents and reading the material before the presentation.*

Tips and Links

Calculation and planning

Project Management Institute: *A Guide to the Project Management Body of Knowledge.* Project Management Institute (Third Edition, 2004)

Berkun, Scott: *The Art of Project Management.* O'Reilly Media (2005)

Best, Kathryn: *Design Management: Managing Design Strategy, Process and Implementation.* AVA Publishing (2006)

FXConverter (Currency Converter): www.oanda.com/convert/classic

Leo Online Dictionaries: www.leo.org

Speaking and writing

Seely, John: *The Oxford Guide to Effective Writing and Speaking.* Oxford University Press (Second Edition, 2005)

Dummy text

www.lorem-ipsum.info/generator3

10.0

The publication of all the information contained in this book was carried out with the greatest possible diligence; however, the possibility of mistakes cannot be ruled out. The authors and the publisher assume no liability for erroneous information or its consequences.

Despite careful checking, we also assume no liability for the content of the links contained in this book. The operators of the websites mentioned are solely responsible for their content. We have no influence on the current or future design or content of the websites mentioned.

Nonetheless, the publisher is grateful for all information about possible mistakes, or further suggestions for improving the book, sent to know-it-all@ die-gestalten.de. In such cases, the publisher will make every effort to correct mistakes, provide clarification and to implement suggestions provided that they are in the public interest.

References

Adobe: www.adobe.com

Adobe Magazine: *Die Technik der Schrift.* 2nd ed. (1995)

Aicher, Otl: *Die Ökonomie des Auges.* Ernst & Sohn, Berlin (1989)

Amman, K.; Kegel, D.; Rausch, B.; Siegmund, A.: *Erfolgreich Präsentieren – Ein Leitfaden für den Seminarkurs.* Landesinstitut für Erziehung und Unterricht, Stuttgart (1999)

Becker, Jochen: *Marketing-Konzeption. Grundlagen des strategischen Marketing-Managements.* Verlag Franz Vahlen, Munich (1993)

Bilz, Silja: *Die Macht der Ziffern – Die visuelle Wirkung von Ziffern und ihre Ausdrucksmöglichkeiten.* Manuscript siljabilz@aol.com (2001)

Birkigt, Klaus; Stadler, Marinus M. et al.: *Corporate Identity.* Verlag Moderne Industrie, Landsberg (2002)

Blana, Hubert: *Die Herstellung. Ein Handbuch für die Gestaltung, Technik und Kalkulation von Buch, Zeitschrift und Zeitung.* K. G. Saur Verlag, Munich (1998)

Bommer, J., o.J.: *Seminar Systemtechnik. Brainstorming, Morphologie, Scenario, Delphi und Delphi-Conference-Methode zum Auffinden und zur Definition von Systemalternativen und zur Erstellung von Prognosen.* Manuscript (2003)

Braehm, H.: Brainfloating. *Neue Methoden der Problemlösung und Ideenfindung.* Wirtschaftsverlag Langen-Müller, Munich (1986)

Bundesverband Digitale Wirtschaft (BVDW) e. V.: www.bvdw.org

Clark, Charles H.: Brainstorming. *The Dynamic New Way to Create Successful Ideas.* Doubleday, New York (1958)

Cornish, Graham P.: *Copyright. Interpreting the law for libraries, archives and information services.* Library Association Publishing, London (1997)

Diezmann, Tanja; Gremmler, Tobias: *Grids for the Dynamic Image.* Ava Publishing, Lausanne (2003)

Diringer, David: *Writing.* Thames & Hudson, London (1962)

Dudenredaktion: *Duden – Die deutsche Rechtschreibung.* Vol. 1., 23rd ed., Dudenverlag; Mannheim, Leipzig, Vienna, Zurich (2004)

Eco, Umberto: *Theory of Semiotics.* Indiana University Press, Bloomington (1979)

Ernst, Bruno: *The Magic Mirror of M. C. Escher. A Revealing Look into the Life and Work of the Most Astonishing Artist of our Time.* Parkwest Publications, Miami (1987)

Flusser, Vilém: *Die Schrift. Hat Schreiben Zukunft?* European Photography, Göttingen (2002)

Forssman, Friedrich; de Jong; Ralf: *Detailtypografie, Nachschlagewerk für alle Fragen zu Schrift und Satz.* Verlag Hermann Schmidt Mainz (2006)

Fraunhofer's Integrated Publication and Information Systems Institute (IPSI): www.ipsi.fraunhofer.de

Frick, Richard; Graber, Christine; Minoretti, Renata; Sommer, Martin: *Satztechnik und Typografie.* Vols. 1 – 5. GDP Verlag, Berne (1989)

Friedl, Friedrich: *Typography. When, Who, How.* Konemann UK Ltd (1998)

Fröhlich, W. D.: *Wörterbuch Psychologie.* Deutscher Taschenbuch Verlag, Munich (2000)

Frutiger, Adrian: *Eine Typografie.* Vogt-Schild-Verlag, Solothurn (1981)

Frutiger, Adrian: *Signs and Symbols. Their Design and Meaning.* Watson-Guptill Publications, New York (1998)

Gesamtverband Kommunikationsagenturen GWA e. V.(Ed.): *Für Studentinnen und Studenten: Einstieg in Kommunikation und Werbung.* Frankfurt am Main (2003)

Gora, Stephan: *Grundkurs Rhetorik. Eine Hinführung zum freien Sprechen.* Klett-Verlag, Stuttgart (1996)

Hochuli, Jost: *Das Detail in der Typografie.* Niggli Verlag, Sulgen (2005)

Homann, Jan-Peter: *Color Management.* Springer, New York (2005)

Hubmann, Heinrich: *Urheber- und Verlagsrecht. Ein Studienbuch.* 8th ed., rev. by Manfred Rehbinder. Beck, Munich (1995)

International Organization for Standardization: www.iso.org

Itten, Johannes: *The Art of Color. The Subjective Experience and Objective Rational of Color.* Wiley, Hoboken (1997)

Jäger, Ludwig: *Ferdinand de Saussure zur Einführung.* Junius-Verlag, Hamburg (2006)

Jensen, Hans: *Die Schrift in Vergangenheit und Gegenwart.* VEB Dt. Verl. d. Wissenschaften, Berlin (1958)

Johansson, Kaj; Lundberg, Peter; Ryberg, Robert: *A guide to Graphic Print Production.* Wiley, Hoboken (2002)

Kaiser, Michael; Liess, Charlotte; Schulz-Neumann, Jörg: *Perspektive als Mittel der Kunst.* Colloquium Verlag, Berlin (1986)

Kapr, Albert: *The Art of Lettering. The History, Anatomy and Aesthetics of the Roman Letter Forms.* K G Saur, New York (1983)

Kapr, Albert; Schiller, Walter: *Gestalt und Funktion der Typografie.* VEB Fachbuchverlag, Leipzig (1977)

Karow, Peter: *Font Technology. Methods and Tools.* Springer Verlag Telos (1994)

Khazaeli, Cyrus Dominik: *Crashkurs Typo und Layout.* Rowohlt Verlag, Reinbek (1997)

Koschnick, Wolfgang J.: *Focus-Lexikon. Werbeplanung, Mediaplanung, Marktforschung, Kommunikationsforschung, Mediaforschung.* Focus Magazin Verlag, Munich (2003)

Koschtial, Ulrike: *Die Einordnung des Designschutzes in das Geschmacksmuster-, Urheber-, Marken- und Patentrecht.* Rhombus-Verlag, Berlin (2003)

Kueppers, Harald: *Basic Law of Color Theory.* Barron's Educational Series, Hauppauge (1981)

Linotype AG: *Zur Geschichte der linearen, serifenlosen Schriften.* Linotype AG

Linotype und Machinery: *A Dictionary of Printing Terms.* Linotype und Machinery, London (1962)

Microsoft: www.microsoft.com/typography

Morrison, R. E.; Inhoff, A. W.: *Visual factors and eye movements in reading.* Visible Language, Providence (1981)

Muzika, Frantisek: *Die schöne Schrift in der Entwicklung des lateinischen Alphabets.* Bd. 2., Dausien, Hanau (1965)

Neutzling, Ulli: *Typo und Layout im Web.* Rowohlt Verlag, Reinbek (2002)

Nimmergut, J.: *Regeln und Training der Ideenfindung.* Wilhelm Heyne Verlag, Munich (1975)

Nitsche, Michael: *Polygraph Dictionary of the Graphics Arts and Communications Technology.* Polygraph Verlag, Frankfurt am Main (1990)

Office for Harmonization in the Internal Market (OHIM): www.oami.europa.eu

Online-Werbeformen:
www.werbeformen.de

O'Reilly, Tim:
www.oreillynet.com/pub/a/oreilly/tim/
news/2005/09/30/what-is-web-20.html

Page: *Service Wortschatz.* Magazine,
8th ed. (1989)

Paulmann, Robert: *double loop – Basiswissen
Corporate Identity.* Verlag Hermann Schmidt
Mainz (2005)

Pohl, R.: *Beobachtungen und Vorschläge
zur Gestaltung und Verwendung von Folien
in Vorträgen.* Psychologische Rundschau,
Münster (1990)

Portal of the European Union:
www.europa.eu.int

Rähm, Klaus u. a.: *TypoGrafik.* H/G Media
Druck & Verlag, Engelsbrand (1996)

Rapp, Christof: Aristoteles. *Rhetorik. Über-
setzung, Einleitung und Kommentar.* 2 Vols.,
Akademische Verlagsgesellschaft Aka,
Berlin (2002)

Reicher, G. M.: *Perceptual recognition as a
function of meaningfulness of stimulus
material.* Journal of Experimental Psychology 81. (1969)

Ritter, M. (Ed.): *Wahrnehmung und visuelles
System.* Spektrum der Wissenschaft-Verlags-
gesellschaft, Heidelberg (1986)

Rostock, University:
www.geoinformatik.uni-rostock.de

Saussure, Ferdinand de: *Course in General
Linguistics.* Open Court, Chicago (1998)

Scheufelen Papierfabrik:
www.scheufelen.com

Schopp, Jürgen F.: *Typografie, Layout
und Translation – von der Bleizeit zum
DTP-Zeitalter.* Universität Tampere,
www.uta.fi/~trjusc

SeminarCentre: *Projektmanagement.* Technology Training. Frankfurt am Main (1995)

Spencer, Herbert: *The visible word.* Royal
College of Art, London (1969)

Stern, Hadley; Lehn, David: *Death of the
Web-Safe Color Palette.* WebMonkey (2000)

Stiebner, Erhardt; Leonard, Walter:
Bruckmann's Handbuch der Schrift.
Verlag F. Bruckmann, Munich (1977)

Suchmaschinenmarketing. Best Practice
Guide Nr. 10, Deutscher Direktmarketing
Verband e. V., Wiesbaden (2005)

Trabant, Jürgen: *Zeichen des Menschen.
Elemente der Semiotik.* fischer perspektiven
Verlag, Frankfurt am Main (1988)

Tschichold, Jan: *Die Bedeutung der Tradition
für die Typografie.* Schriften, Berlin (1964)

Tschichold, Jan: *Formenwandlung der Et-
Zeichen.* D. Stempel AG, Frankfurt am Main
(1953)

Tschichold, Jan: *Treasury of Alphabets and Lettering.* W. W. Norton & Company, New York (1995)

Type Directors Club (Ed.): Nachdruck (1986) *Elementare Typografie.* Sonderheft der Typografische Mitteilungen 1925 – Reprint. Verlag Hermann Schmidt Mainz

Unicode: www.unicode.org

Vöhringer, Karl: *Druckschriften – kennen-lernen, unterscheiden, anwenden.* Verlag Form und Technik, Stuttgart (1989)

W3 Consortium: www.w3.org

Waldmann, Werner; Zerbst, Marion: *DuMont's Handbuch Zeichen und Symbole. Herkunft – Bedeutung – Verwendung.* DuMont Literatur und Kunst Verlag, Cologne (2006)

Weidemann, Kurt: *Wo der Buchstabe das Wort führt. Ansichten über Schrift und Typographie.* Hatje Cantz, Ostfildern (1994)

Wendt, Dirk: *Lesbarkeit von Druckschriften, Lesen Erkennen.* TGM, Munich (2000)

Willberg, Hans Peter; Forssman, Friedrich: *Erste Hilfe in Typografie.* Verlag Hermann Schmidt Mainz (2001)

Zimbardo, Philip; Gerrig, Richard: *Psychology and Life.* Allyn & Bacon, Boston (2001)

Zurich, University: www.unizh.ch/RZU/publications/ps/truetyed.ht

Index

Symbols
1080i 120, 121
16:9 picture format 121
16 bit colour depth 14
2-byte character coding 91
2D grid 131
4 Ps 218
720p 120, 121
7 bit binary code 92
80:20 principle 285

A
AC-3 file 123
accentuation 78
acceptance 250
accessibility 141
achromatic 12, 13, 29
    colour reproduction 178
    degree 29
Active Server Page 141
ad 221, 230
additive colour mixing 9, 10, 12
additive colours 9
additive primary colours 9
Adobe After Effects 126
Adobe Illustrator 104
Adobe InDesign 106
Adobe Photoshop 108
Adobe PostScript Type 1 90
advertisement 221, 230
advertising 221, 239
    advertising medium 222
    banner 221
    basic model for effective advertising 221
    comparative advertising 276
    conducting an advertising campaign 224
    disciplines in 226
    economic aims of 221
    internet advertising 277
    non-economic aims of 221
    online advertising 224
    pop-ups 221
    resources 222
aerial perspective 20
agreement
    verbal 250
    written 250
ai 104
AIDA formula 221
aif 104
aims, planning jobs and projects 284
aims of advertising
    economic 221
    non-economic 221
AJAX, Asynchronous Javascript and XML 117
Akzidenz Grotesk 65, 66
Alberti, Leon Battista 17
alinéa 83
alphabet 63, 68
altar folding 199
AMA, American Marketing Association 218, 239
ambiguous pattern 33
American Marketing Association 218, 239
American Standard Code for Information Interchange 92
ampersand 69
amplitude modulated screen 173
AM screen, amplitude modulated screen 173
analogue proofing process 181
anchor point 90
animated banner 224
animation 132
    animating manually 132
    behaviour and sets of rules 130
anti-aliasing 94, 134
APIs 117

apostrophe 80
approval
   print approval 204
arch 68
area 26
Aristotle 298
arm 68
arrow 26
art buyer 227
art director 226
artwork 176, 182
ascender 68, 85
ASCII, American Standard Code for
Information Interchange 92
A series 152, 154, 155
   2A0 154
   4A0 154
   A0 154
   A1 154
   A10 154
   A2 154
   A3 154
   A4 154
   A5 154
   A6 154
   A7 154
   A8 154
   A9 154
ash content 159
ASP, Active Server Page 141
Association Typographique Internationale
72
asymmetry 30
Asynchronous Javascript and XML 118
ATypI, Association Typographique
Internationale 72
authoring 123
AutoCAD 104
autotype screen 173
avi 104
axial setting 77

B
back cover 202
back insert 202
ball 68
band, decorative 202
banner 224, 225
   animated 224
   Full Banner 225
   interactive 225
   Medium Rectangle 225
   Pop-Up 225
   Standard Skyscraper 225
   static 224
   Super Banner 225
bar, letter anatomy 68
bar chart 303, 304
barrier-free 141, 142
   basic rules for barrier-free access 142
base area 20
base line 20, 68
base line grid 43
basic colours 10
basic elements 25
basic grid 129
Bauermeister, Benjamin 73
beating, paper fibre
   free 157
   wet 157
Beta SP 120
Bézier curves 90
binary digit 111
binding processes 201
   block gluing 201
   block stapling 201
   fan binding 201
   perfect binding 201
   plastic binding 201
   spiral binding 201
   stapled binding 201
   stitched binding 201
   threadless binding 201

wire-o binding 201
    wire binding 201
binocularity 15
bird's-eye view 19
bit depth 111
bitmap 110
    bitmap fonts 90
bit rate 124
Black Letters and Broken 74
blank 81
bleed 177
blind stamping 199
block
    gluing 201
    stapling 201
    thickness 161
block, print forms 183
    halftone block 183
    line block 183
Blue-Ray 122, 124
blueprint 180
bmp 104
board 162
    Bristol 162
    chromium sulphate board 163
    Chromolux 163
    duplex 163
Bodoni, Giambattista 65
body 57
body language 298, 300
body text 61
boilerplate 236
book cover 202
bookmark, browser 115
bookmark, ribbon 202
books on demand 196
bowl 68
brackets 80
brainstorming 45
briefing 41, 219
    elements of an agency briefing 219

Brin, Sergy 229
British Standards for Type Classification 73
broadsheet 155
Brunelleschi, Filippo 18
BS-5261 2 167
B series 152, 154
    B0 154
    B1 154
    B10 154
    B2 154
    B3 154
    B4 154
    B5 154
    B6 154
    B7 154
    B8 154
    B9 154
bullet point manuscript 300
button 135
    call-me 230
byte 111

C
calcium carbonate 157
calender 158
calendering 158
call-me button 230
cap height 68
Capitalis Monumentalis 63
Capitalis Quadrata 63
cardboard 162
Caroline Minuscule 63, 64
Cascading Style Sheets 135
case law 249
cast-coated paper 162
cavalier perspective 22
CD manual 238
cellulose 156
central perspective 18
    golden rules of 19
central unit 112

centre
   mathematical 26
   optical 26
centred setting 76
channel
   sender-recipient model 222
channels of communication 221
chapter number 82
character
   character set 69, 70, 86
   character spacing 75
   numeric character 69
   special characters 69
   width of a character 75
character coding 92
   ASCII 92
   Multiple Master 92
   unicode character coding 91
characteristic printing curve 188
characteristic style of a character 62
character set, HTML 138
chart 301, 303
   bar chart 303, 304
   digital presentation chart 301
   pie chart 303
checklist
   artwork 182
   choosing paper 166
   colour climate 13
   creating a presentation 297
   design grid 43
   design resources 30
   elements of an agency briefing 220
   perspective construction 23
   planning process 289
   planning schedule for a presentation 300
   PR concept content 236
   presentation 304
   project monitoring 295
   project planning phases 294

   what must planner and procedures be
   able to deliver 288
chromatic
   colour reproduction 178
   degree 29
   quality 29
Cicero 57
CI manual 40, 237
circle 26
ClearType 94
client, payment by 253
Cline, Craig 116
CMS, Content Management System 133
CMYK 186
CMYK working colour models 169, 170
coated paper 161, 170
coating 158, 161
   coating machine 158
   colour coating 158
codec 123, 124
cold light 10
coldset inks 187
collecting societies 266, 280
colon 81
colour 8, 140
   colour contrasts 28
   colour density 188
   colour depth 14
   colour effects 11
   colour perspective 19
   colour psychology 11
   colour reproduction 178
   colour scale 180
   colour separation 178
   colour set 178
   colour settings 171, 172
   colour stimulus 8
   colour symbolism 11, 12, 13
   colour theory 10
   colour wedge 188
   perceiving colour 8

sensitivity to colour 8
coloured edge 202
colour management 169, 207
    workflow 171, 172
colour profiles 169, 172
    AdobeRGB 169
    CMYK working colour model 169
    input profile 169
    LAB colour model 169
    monitor profile 170
    printer profile 170
    RGB working colour model 169
    sRGB 169
colours
    additive colour mixing 10
    additive colours 9
    additive primary colours 9
    basic colours 10
    black 13
    blue 8, 9, 10, 12, 14, 29
    cold colours 20, 28
    colours on the web 14
    complementary colours 11, 29
    cyan-blue 11
    elementary colours 11
    green 8, 9, 10, 11, 13, 14, 28, 29
    grey 13
    light colours 9
    magenta 12
    magenta-red 11
    non-luminous colours 9, 10
    of the spectrum 8
    orange 12
    orange-red 11
    Pantone colours 186
    primary colours 10
    printing inks 186
    RAL colours 186
    red 8, 9, 10, 12, 14, 29
    RGB colours 9, 14
    secondary colours 9, 11
    special colours 186
    spot colours 186
    subtractive colour mixing 10
    subtractive colours 9
    tertiary colours 11
    violet 12
    violet-blue 11
    warm colours 20, 29
    white 12, 28
    yellow 9, 10, 11, 12, 28, 29
column title 82
commissioned orders 245
commissioning work 250
common law trademarks 273
communication
    channels of 221
    in project management 294
    policy 218
competition law 268, 275
    Unfair Competition Act 268, 276
complementary colours 11, 29
compositing 125
compression 110, 124
computational design 127, 128
computer 112
computer-to-film 184, 185
computer-to-plate 184, 185
computer-to-print 184, 185
computer network 114
concept 287
concept sign 36, 63
concertina folding 199
concession of rights of use 263
    basic rights of use 264
    conditional 264
    exclusive rights of use 264
    unconditional 264
cones 8
container formats 103
Content Management System 133
continental copyright system 256

continuous text 61
contract 245
   contract for labour and services 248
   licence agreement 266
   service contract 248
   utility contract 266
contract for labour and services 248
contrast, colour 28, 29
   chromatic contrast 28
   colour-as-such contrast 28
   complementary contrast 29
   light-dark contrast 28
   quality contrast 29
   quantity contrast 29
   simultaneous contrast 29
   warm-cold contrast 28
contrast, layout 29, 30
   area contrast 30
   colour contrast 30
   formal contrast 29
   order contrast 30
   size contrast 30
   thickness contrast 29
control line 181, 182
control wedge 181, 182
   Fogra CMYK media wedge 182
conventional screen 173
cookies 115
copperplate engraving 194
copy protection 123
copyright
   act 258, 261
   infringement of 263
   notice 261, 262
   personal 267
   registering 260
   regulation, EU 257
copyright law 254, 255, 267
   French 258
   German 256, 259
   UK 262
   US 265
copyright office 258, 279
   register with 261
copyright protection
   Europe 257
   on industrial design 271
copyright system
   Anglo-Saxon 258
   continental 256
corporate design 238
   manual 237, 238
corporate identity 237, 238
   manual 237
   phases 238
correction marks 167, 168
   British standard, BS-5261 2 167
   German standard, DIN 16511 167
   international standard, ISO 5776 167
corrections by authors 249
cost estimate 247
cost per order 231
counter 68
CPM, cost per mille 222
CPT, cost per thousand 222
CPU, Central Processing Unit 112
creasing 200
creative director 226
Creatives 226
creative techniques 44
creativity 44
crib-sheet method 299
crib methods 300
crisis management 233
cross folding 199
cross media 184
CRT screen 112
C series 154
CSS, Cascading Style Sheets 135, 137, 139, 142
CtF, computer-to-film 184, 185
CtP, computer-to-plate 184, 185

cuneiform 63
cut-out 177
cutting waste, paper 156
cylinder press 191

D
damage, defaulted payment 253
dash 80
database marketing 231
data format 102
daylight 10
dcs 104
dd, Didot point 57
Decorative and Display 74
decorative band 202
degree symbol 81
densitometer 188
descender 68, 85
design
   fundamentals of 25
   golden rules of 48
   laws of 31
   notice 271
   patent law 270
   protection 254
designing a presentation 301
desktop computer-to-film 184
devices
   input 112
   output 112
diagonal perspective 19
dialogue marketing 230
diametric perspective 22
Didones 73
Didot family 65
Didot point 57
Digi-Beta 120
digital presentation charts 301
digital printing 189, 195
digital proof 181
dimensional stability 158

dimension sheet 161
DIN, German Institute for Standardization 72, 152
DIN 16511 167
DIN 16518 72
Direct Imaging 184
directive on electronic commerce 277
direct marketing 230
disciplines in advertising 226
displacing 130
   displacement map 130
display, three-dimensional 16
display face 62
distance 20
distribution rights 263
dithering 14
DLT master cartridge 123
DNS, Domain Name System 114
doc 104
doctor blade 193, 194
Doherty, Dale 116
Dolby Digital Plus 124
Domain Name System 114
dot 25, 173
dot gain 189
D series 154
DTP, desktop publishing 66
DTP point 57
DTS-HD 124
DTS file 123
ductus 62
dummy 41, 161
Dürer, Albrecht 18
DV Cam 120
DVCPro25 120
DVCPro50 120, 123
DVD 119, 123, 124, 125, 146
DVD authoring 125
dwg 104
dye 157, 165
dye-cutting 197, 202

dynamic grid 129
dynamic structure 131

E
e-commerce 277
e-mail 114
e-marketing 228
ear 68
ECI, European Color Initiative 170, 207
Eco, Umberto 35
edge, coloured 202
editing 125
effect 132
    kinetic effect 33
    screen 174
    three-dimensional effect 16
    varnish 197
Egyptian faces 65
electro-photographic printing processes
195
electromagnetic waves 8
electronic paper 165
Elemental Typography 66
elementary colours 11
elements, basic 25
ellipse 27
emblem 36
embossing 199
emphases 137
empty space 81
em space 83
emulsion varnish 197
encapsulated 110
encoding 125
endpapers 202
end tag 136
English paperback binding 202
eps 104
Escher, Maurits Cornelis 15, 33
E series 154
etching 194

Ethernet 114
EU copyright regulation 257
European Color Initiative 170, 207
European scale, printing inks 186
event manager 227
exchange format 102
exclamation mark 81
exe 104
executive, paper format 155
exhibition rights 263
exploitation rights 248, 263
exposure 180, 181, 185
Extensible Markup Language 139
extraneous costs 247, 252
extranet 115
eye, the human
    cones 8
    retina 8
    rods 8
eye level 20

F
fair copy 41
fan binding 201
feeding marks 177
feeds 116
    RSS 116
fh 104
Fibonacci series 42
figure-ground 33
    distinction 33
file cards 301
file formats 102, 104, 105, 106, 107, 108,
109, 110
File Transport Protocol 114
fillers, paper 157
Final Cut Pro 126
final draft 41
final project report 295
financial and investor relations 233
finishing 197

fisheye projection 23
fla 104
flashes 176
flash plug-ins 124
flat-rate payment 247
flat screen 112
flexo printing 192
flow plan 292
flush left setting 76
flush right setting 76
flv 104
FM screen, frequency
modulated screen 174
fold 199
folding 199
    altar folding 199
    concertina folding 199
    cross folding 199
    parallel folding 199
    single folding 199
    zigzag folding 199
folio head 82
font
    bitmap font 90
    fonts on a display screen 93
    font technology 90
    web design 134, 137, 140
font, HTML tag 137
font formats 90
    OpenType 91
    PostScript 90
    TrueType 90
footer 82
footnote 59, 82
    footnote symbol 81
formats
    harmonious page formats 42
formats, standard 42
formatted paper 156
Fourdrinier wire paper machine 157
fps 121

fps, frames per second 119
fractalisation 38
fragrant varnish 198
frames per second 119
free beating 157
French paperback binding 202
frequency modulated screen 174
front cover 202
front cut 200
front view 161
Frutiger, Adrian 25, 66, 84
FTP, File Transport Protocol 114
Full Banner 225, 226
full frame 120
further expenses 247
Futura 66

G

game design 127
Gantt, Henry Laurence 293
    Gantt method 293
Garamond, Claude 64
GCR, Gray Component Replacement 179
Geml, Richard 221
general laws of protection 255
General Terms and Conditions
of Business 245
Geraldes 73
Gestalt psychology 31
Gestalt theory 31
gif 104
Gill 66
Gillam, Barbara 24
glance over the shoulder 219
glyphs 67
    glyphs palette 91
Goethe, Johann Wolfgang von 10, 29
golden section 42, 82
Google 226, 228, 229, 239
Gothic Minuscule 63, 64
GRACoL 170, 181

grain 159, 160
grain direction test methods 160
    bending test 160
    fingernail test 160
    moisture test 160
    tearing test 160
graphic designer 226
graphics card 111
graphics file 134
gravure printing 193
grey balance 188
grey value 77, 93
grey wedge 188
grid 40, 129, 131
    2D grid 131
    base line grid 43
    basic grid 129
    construction grid 42
    design grid 42
    dynamic grid 129
grid systems for digital and
dynamic media 129
grotesque 65
ground plan 20
GTCB, General Terms and
Conditions of Business 245, 253
Guilford, Joy Paul 44
guillemet 80
Gutenberg, Johannes 64
Gutenberg Bible 64
gutter 82

H
H.264 124
hairline 64, 68
halftone 172, 173
    dots 173
    element 173
    graphic 173
    image 173
    screen 173

halftone blocks 183
Hamburg Model 294
handouts 304
Handwritten 74
hardback 202
hardboard 162
hardcover 202
hard proof 182
hardware 111, 112
    central unit 112
    computer 112
    CPU 112
    peripherals 112
    RAM 112
harmonious page formats 42
harmony 17
hatching 26
HD, High Definition 120, 121
HD-DVD 122, 124
HD material, reproducing 122
HD productions, the advantages of 122
HD ready 122
HD recording formats 123
    AVC 123
    DVCProHD 123
    HD Cam 123
    HDV 123
HDTV 120
    1080i 120, 121
    720p 120, 121
header 82
headings 59, 61, 62, 82, 137
headline typeface 62
heatset inks 187
Helvetica 66, 70
Hermann, Caspar 189
hexadecimal 14, 140
hieroglyphics 36, 63
High Definition 120, 121
hinting 90, 95
hints 90

HKS colour system 186
homepage 133, 144
horizon 18
   horizon line 20
hot foil stamping 198
hot metal setting 57
hourly wage 247
hqx 104
htm 104
HTML, Hypertext Markup Language
104, 135, 136, 137, 138, 139, 141
   basic framework of an HTML file 138
   body 139
   head 138
   special HTML characters 137
   tags 137
HTTP, Hypertext Transfer Protocol 115
Humanes 73
hybrid systems 72
hydrophilic 189
hydrophobic 189
hyperlinks 136
hypertext 136
Hypertext Transfer Protocol 115
hyphen 80

I

IAB, Interactive Advertising Bureau 225,
239
icon 35
idea
   idea book 44
   imitation of an 268
   protection of 268
ideal proportion 42
ideogram 36, 63
iff 106
IFRA 170
illusions 24
   visual perception phenomena 84
image file 134

imposition 179
   form 179
   scheme 179
imprint obligation 277
in, inch 57
in-site format 226
inch 57
incidental technical expenses 247
Incises 73
ind 106
indent 83
index 35
indirect printing process 190
industrial design
   copyright protection on 271
industrial property 270
information architecture 126
infringement, of copyright 263
initials 79
ink absorption 188
inkjet printing 196
input device 112
instant messaging 115
intelligent web 118
interaction 126, 127, 131
   design 126
interactive
   banner 225
   media environments and
   installations 127
   movement 131
   structure 131
Interactive Advertising Bureau 225, 239
interchangeable file format 106
interface 126, 128, 131, 143
   design 126
interim reports 295
interlaced 119, 120
internal communications 232
International Organization for
Standardization 152, 170, 207

International Press Telecommunication Council 267, 280
internet 114, 146
   advertising 277
   trends and developments 116
inverted commas 80
investor relations 233
invoice 251
invoice, paper format 155
IP address 114
IPTC, International Press Telecommunication Council 267, 280
IR radiation 8
ISO, International Organization for Standardization 152, 170, 207
ISO/DIN series 152, 153, 154
   A series 154
   B series 154
   C series 154
   D series 154
   E series 154
ISO 12647 166, 169, 170, 186
ISO 2846 186
ISO 5776 167
ISO 9541 90
isometric perspective 22
issues management 232
italics 79, 89
Itten, Johannes 10, 28

J
Java 141
JavaScript 141
jfif 106
jpeg 106
   compressing a 110
jump-line structure 144
justified setting 77

K
kaolin 157
Kapr, Albert 27
kern 68
kerning 86, 87, 89
   aesthetics tables 89
   pair 89
   tables 89
   value 89
key frames 130, 132
keyword 228, 229
kinetic effect 33
Kodak PhotoCD 106
Küppers, Harald 9, 10, 11, 29, 186

L
L/cm 173
lamination 198
LAN, Local Area Network 114
landscape format 160
language 34
Lanham Act 273
laser printing 196
Lasswell, Harold D. 222
laws of design 31
   law of closure 32
   law of common fate 32
   law of continuation 32
   law of experience 33
   law of proximity 31
   law of similarity 32
   law of simplicity 31
   law of symmetry 32
   laws of Gestalt 31
layout 41
   file formats 102
   page layout 81
   process 41
   proof 181
   rough layout 41
LCD screen 112

leading 75
leg 68
legal, paper format 155
legal rule of thumb 249
letter, paper format 155
letterpress printing 191
letters 67
  appearance of 84
  basic framework of 85
  capital 67
  illuminated 79
  inner structure of 60
  letter anatomy 68
  small 67
  weight of 85
  width of a letter 85
Lewis, E. St. Elmo 221
liability risks 246
licence, non-exclusive 266
licence, picture 41
licence agreement 266
licencee 264, 266
ligatures 69, 71
light 8
  cold light 10
  colours 9
  daylight 10
  IR radiation 8
  light and shadow 16
  light temperature 10
  UV radiation 8
  warm light 10
line 25
Linéales 73
linear
  method 300
  perspective 18
  structure 144
line blocks 183
line length 76
linen tester 187

lines per centimetre 173
lines per inch 173
link, letter anatomy 68
link popularity 229
lipophilic 189
lipophobic 189
litho 191
lithography 65, 190
Local Area Network 114
logo 37
  constructing a logo 40
  designing a logo 38
  protection zone 40
  trends and development 37, 38, 39
lowercase 63, 64, 67, 71
LPCM 124
lpi 173
lzw 106

M
machine proof 180
Macromedia Freehand 104
macrotypography 56
Madrid Trademark Agreement 275
Maeda, John 128
mailing 231
main stroke 64, 68
majuscules 67
Manuaires 73
marginalia 59, 82
margins 82
mark 37
marketing 218
  communication 218, 232
  social 38
marketing mix 218
  4 Ps 218
mashup 117
Massachusetts Institute of
Technology 128, 165
mathematical centre 26

mathematical signs 80
measurement systems 57
Mécanes 73
media planner 226
media relations 232
Medium Rectangle 225
meta-search engines 228
meta-tags 228
metafile formats 110
Microsoft Word 104
microtypography 56
mid 106
Miedinger, Max 66
military perspective 22
millimetre 57
mind mapping 46
Mini-DV 120
minuscules 67
misregistration 187
MIT, Massachusetts Institute of
Technology 128, 165
MLP 124
mm, millimetre 57
MM, Multiple Master 92
mobile application design 127
model
    Open Source 117
    sender-recipient 222
Modern Face 74
moiré 174
monitor 112, 133
    typography 94
Morris, Charles W. 35
motion
    design 126
    graphics 132
    tracking 129
mov 106
movable type 64
moving pictures 119
mp3 106

mp4 106
MPEG 124
MPEG2 123, 124
mpv 106
MTA, Madrid Trademark Agreement 275
Mueller-Lyer, F.C. 24
Multiple Master 92
multiplexing 123

N
narrow web 159
navigable structure 131
navigation 135
NDA, non-disclosure and
confidentiality agreement 269
Necker cube 33
net structure 145, 300
network types 114
    Ethernet 114
    extranet 115
    internet 114
    LAN 114
    VPN 115
    WAN 114
    WLAN 114
Newton, Isaac 10
Nielsen, Jakob 142
non-disclosure and confidentiality
agreement 269
non-luminous colours 9, 10
NTSC, National Television Systems
Committee 119, 120, 123
number 69
    chapter number 82
    page number 82
numerals 64
    arabic numerals 64, 69
    lowercase numerals 81
    normal numerals 81
    small caps numerals 81
    table numerals 81

O

O'Reilly, Tim 116
object tracking 130
offerer identification 277
offer of services 246
Office for Harmonization in the Internal Market 270, 274, 279
offset printing 189, 190
Ogilvy, David 219
OHIM, Office for Harmonization in the Internal Market 270, 274, 279
Old Face 64, 74
omission marks 80
on-site format 226
online advertising 224
Online PR 234
opacity 157
Open Source Model 117
OpenType 91
operating system 113
optical
    centre 26
    illusion 24
    sampling 187
optomechanical typesetting system 66
organigram 303
orphan 83
Osborn, Alex F. 45
otf 106
output
    device 112
    medium 129
    process 180
overhang 68
overlapping 176
overprinting 176, 177
overshoot 68

P

pad printing 194
page
    design 41
    layout 81
    numbers 82
Page, Larry 229
PageRank 229
page size, web design 133
PAL, Phase Alternating Line 119, 120, 123
palette
    standard palette 14
    web-safe palette 14
PAL TV 119
PANOSE System 73
Pantone colours 186
paper
    electronic 165
    formatted 156
    manufacture 156
    qualities 156
    reel 156
    sheet 156, 157
    stock 157
    surface 158
    thickness 160
    volume 160
paper, special 163
    carbon 163
    carbon-copy 163
    chromium 164
    coloured 164
    fine 163
    high gloss 164
    Japan 164
    medium-fine 164
    parchment/greaseproof 163
    pergamin 163
    vat 164
    WP 163
paperback 202

paperback binding 202
  English 202
  French 202
  stiff 202
  Swiss 202
paper by use 164
  bulky paper 164
  inkjet paper 164
  laser printing paper 164
  low grammage paper 164
  LWC paper 165
  magazine paper 164
  offset paper 164
  poster paper 165
paper fibre 157, 159
  free beating of the 157
  wet beating of the 157
paper finishing 158
  calendering 158
  coating 158
  stamping 158
paper formats 152
  Canadian 154, 155
  ISO/DIN 152
  ISO/DIN RA 155
  untrimmed 155
  US 154, 155
paper types 156, 166
  art paper 162
  basic machined paper 158
  cast-coated paper 162
  coated paper 161, 170
  glossy coated 166
  illustration printing paper 162
  LWC paper 161, 165, 166, 170
  matt coated 166
  newsprint 166
  rag-containing paper 156
  uncoated paper 161, 166, 170
  wood-containing paper 156
  woodfree paper 156

paper volume 161
paper weight 160
  paper grammage 160
papyrus 156
parallel folding 199
parallel perspective 21
parametric transfer 129
pasteboard 162
  machine pasteboard 162
patent 254, 271, 272, 279
pattern, ambiguous 33
pattern design, registration of a 257, 270
pay-per-click 229
payment
  by client 253
  deadline 252
  defaulted 253
  for services 247
  terms of 248
pcd 106
PCM format 123
pct 106
PDA, Personal Digital Assistant 127
PDF, Portable Document Format 106, 179,
184, 203
PDF/X 179
Peirce, Charles S. 35
per cent sign 81
  one tenth of one per cent sign 81
perception 60
perfect binding 201
perforating 202
pergamin 163
period of protection 257
  EU trademark 274
  international trademark 275
  national trademark (German) 273
peripherals 112
permalinks 117
per procurationem 246
Personal Digital Assistant 127

personalised printing 195, 196
perspective 15
    aerial perspective 20
    bird's-eye view 19
    cavalier perspective 22
    central perspective 18
    colour perspective 19
    diagonal perspective 19
    diametric perspective 22
    fisheye projection 23
    history of perspective 17
    isometric perspective 22
    linear perspective 18
    military perspective 22
    parallel perspective 21
    significance perspective 23
    worm's-eye view 19
photo agencies 268
photopolymer letterpress plates 192
photosetting 66
PHP Hypertext Processor 141
physical inks 187
picking resistance 159
pictogram 36, 63
pictorial script 36
pictorial signs 63
picture agencies 41
picture archives 268
picture plane 20
pie chart 303
pilcrow 83
pitch 268
pixel 93
    counts 229
    file formats 110
planning
    a presentation 300
    four golden rules of 287
    jobs 284
    jobs and projects 284
    projects 287

plastic binding 201
platen press 191
plates
    photopolymer letterpress plates 192
    rubber plates 192
plots 182, 203
png 106
point 57
    Didot point 57
    DTP point 57
    typographical point 57
point-of-sale 222
Pop-Up 225, 226
portrait format 160
PostScript 57, 90, 106
PostScript Type 1 90, 93
ppt 106
PR, public relations 232, 233, 234, 236, 239
pragmatics 34
PR concept 236
pre-flight 180
pre-press 167
pre-print 167
presentation 297, 298, 300, 304
    communicative elements in a 298
    creating a 297
    designing a 301
    methods 303
    planning a 300
press release 234
    structuring a 235
press section 234
prestigious finishing 202
    dye-cutting 202
    laminating 198
    perforating 202
    punching 202
    varnishing 197
price comparisons 276
primary colours 10
print approval 204

printing 186
  first run and back-up printing 187
  four-colour printing 186
  seven-colour printing 186
printing curve, characteristic 188
printing inks 186
  coldset inks 187
  European scale 186
  heatset inks 187
  Pantone colours 186
  physical inks 187
  process colours 180, 186
  RAL colours 186
  special colours 186
  spot colours 186
  standard colours 186
  structural inks 187
  UV inks 187
printing on demand 195, 196
printing position 177
printing problems 205
  channel stripes 206
  defective blanket 205
  doubling 206
  hickeys 205
  ink stripes 206
  mechanical flaws 206
  smears 206
printing process, indirect 190
printing process control 187
printing processes 189
  copperplate engraving 194
  cylinder press 191
  digital printing 195
  electro-photographic printing 195
  etching 194
  flexo printing 192
  gravure printing 193
  indirect printing process 190
  inkjet printing 196
  laser printing 196

  letterpress printing 191
  lithography 190
  offset printing 189, 190
  pad printing 194
  personalised printing 195, 196
  printing on demand 196
  recess printing 194
  rotary printing 191
  screen printing 194
  serigraphy 194
  UV printing 190
  waterless offset printing 190
print varnish 197
process colours 180, 186
production, commercial practice 203
production manager 226
product publicity 232
programming 114, 135, 146
  languages 128
progressive 120
project diary 295
project management,
communication in 294
project plan 289, 290, 291
  list of milestones 291
  project approval 291
  project budget 291
  project definition 291
  project variables 291
  supplementary plans 291
  types of plan 292
project planning process 289
project report, final 295
project structure 290
  plan 290
proof 180
  analogue proofing process 181
  correction marks 167, 168
  digital proof 181
  hard proof 182
  layout proof 181

machine proof 180
soft proof 182
proofing system 181
proportion, ideal 42
prospektiva 15
protection
    general laws of 255
    of ideas 268
    of printed characters 271
    of titles 275
    rights for industrial property 270
ps 106
psd 108
pt, DTP points 57
public affairs 232
public relations 232, 233, 234, 236, 239
    functions of 233
    instruments of 233
pulp 156, 157, 159, 164
punch 57
punching 197, 198, 202
punctuation 69, 71
    marks 81

Q

Quark Xpress 108
question mark 81
Quicktime 119, 124, 125
QuickTime Movie 106
quotation marks 80
qxd 108

R

R, registered 272
ragged setting 77
RAL colours 186
RAM 112
ranking 228
rar 108
Raster Image Processor 183
RBC, Revised Berne Convention 255

re-briefing 219
reading 60
    behaviour 60
    habits 25
Réales 73
Really Simple Syndication 116
recess printing 194
recipient
    sender-recipient model 222
recordation 266
recording formats, SD 120
rectangle 27
Rectangle, banner 225
reel 156
Reeves, Rosser 219, 222, 223
register
    correct register 43
    keeping to register 43
    register marks 177
registered 272
registered design, Europe 270, 274
registered design law 254, 270
registration, pattern design 257, 270
regularity 17
relief stamping 199
reminder system, planning 285
Renaissance, Italian 17
Renner, Paul 10
reproduction 167
    control 179
resolution, screen 113
response element 230
response quota 231
reticent typefaces 63
retina 8
Revised Berne Convention 255
RGB 9
RGB working colour models 169
rhetoric 297, 298
    crib-sheet method 299
    golden rules of 298

rhythm 17, 30
ribbon bookmark 202
right of reproduction 263
rights-managed 268
rights of use 248
     basic 264
     consessions of 263, 264
     exclusive 264
RIP, Raster Image Processor 183, 184, 185, 203
robots 228
rods 8
Roman capitals 64
Roman typeface 64
rotary printing 191
rough layout 41
royalty-free 268
RSS feeds 116
rtf 108
rubber plates 192
Rubel, Ira W. 189
Rubin, Edgar J. 33
running head 82

S

saccades 60, 61
sales promotion 218
sample plans 296
sans-serif 65, 74
satinising 158
Saussure, Ferdinand de 35
schedule 303
Schiller, Walter 27
Schulz von Thun, Friedemann 294
     Hamburg Model 294
scoring 200
screen
     angle 175
     frequency 173, 189
     printing 194
     resolution 113

     screen angle meter 189
     size 113
     wedge 189
screen, monitor 112, 113, 172
     CRT 112
     flat 112
     LCD 112, 122
     plasma 122
     resolution 113
     size 113
screen types 173
     AM screen 173
     effect screen 174
     FM screen 174
     halftone screen 172, 173
script 60
     anatomy of a script 60
     German Gothic script 64
     hieroglyphic 36
     pictorial 36
Scriptes 73
Scripts 74
     Foreign Scripts 74
SD, Standard Definition 119
     recording formats 120
SDTV 120
     NTSC 119, 120, 123
     PAL 119, 120, 123
search engine marketing 228
SECAM 119
secondary colours 9, 11
second generation web 116
selective binding 202
self-discipline, planning 285
semantics 34, 60
semicolon 81
semiotics 34
sender-recipient model 222
Senefelder, Alois 190
separation 178
serifs 68

serigraphy 194
service contract 248
setting
    leaded setting 76
    solid setting 76
setting alignment 76
    axial setting 77
    centred 76
    flush left 76
    flush right 76
    justified setting 77
    ragged setting 77
    symmetrical setting 76
setting priorities, planning 285
set width 75, 94
SGML, Standard Generalized Markup
Language 139
shadow 16
    types of shadow 16
shank 57
sheer plan 20
sheet 156
    making 157
    signatures 177
shoulder 57, 68, 75
sieve section 157
sign 34
    concept sign 63
    pictorial sign 63
    sign theory 34
    types of signs 35
signet 36
signifiant 35
significance perspective 23
signifié 35
signified 35
signifier 35
single folding 199
single frame structure 145
sit 108
sitemap 143

size, paper production 157, 159, 163
size differences in three
dimensional display 17
sizes, type 59
sketch 41
sketchbook 44
Slab-Serif 74
slab-serif faces 65
slash 81
small caps 79
social marketing 38
social software 118
social web 118
soft proof 182
software 111, 113, 125
    compositing 125
    DVD authoring 125
    editing 125
    encoding 125
    operating system 113
source code 135
spacing 75, 86, 87
    basic spacing 88
    line spacing 75
    word spacing 75
spatial quality 15
special colours 186
spiders 228
spiral 26
    binding 201
spoilage 196
sponsored links 229
spot colours 186
square 27
stamping 158
standard colours 186
Standard Definition 119
standard formats 42
Standard Generalized Markup
Language 139
Standard Skyscraper 225

standpoint 20
stapled binding 201
static banner 224
stem 68
stiff paperback binding 202
stimulus, colour 8
stitched binding 201
stock vat 157
stress 68
stroke 26
structural inks 187
structure 26, 31, 144
  jump-line 144
  linear 144
  net 145
  single frame 145
  tree 144
style guide 238
subscription services 116
subtractive colour 9
subtractive colour mixing 10
suffix 102
  ai 104
  aif 104
  avi 104
  bmp 104
  dcs 104
  doc 104
  dwg 104
  eps 104
  exe 104
  fh6 104
  fla 104
  flv 104
  gif 104
  hqx 104
  htm 104
  html 104
  iff 106
  indd 106
  jfif 106
  jpg 106
  lzw 106
  mid 106
  mov 106
  mp3 106
  mp4 106
  mpv 106
  otf 106
  pcd 106
  pdf 106
  pict 106
  png 106
  ppt 106
  ps 106
  psd 108
  qxd 108
  rar 108
  rtf 108
  sit 108
  swa 108
  swf 108
  tif 108
  tiff 108
  txt 108
  wav 108
  wmf 108
  xls 108
  xlsx 108
  zip 108
Super Banner 225
surface finishing 197
swa 108
swf 108
Swiss paperback binding 202
SWOP 170, 172, 179
SWOT analysis 236
symbols 34, 35, 36
symmetrical setting 76
symmetry 17, 30, 32
  typeface design 85
syntax 34

T

table, HTML tag 137
tabloid, paper format 155
tagging 117
tags 117, 136, 137, 138, 139
    end tag 136
    font tag 137
    HTML tags 137
tax identification number 252, 277
TCP/IP, Transmission Control Protocol/
Internet Protocol 114
terminal 68
territoriality principle 255
tertiary colours 11
testimonial 223
text
    body text 61
    continuous text 61
text editors 128, 135, 136
text file format 104
Textura 64
Thiéry figure 33
third party services 246
threadless binding 201
three-dimensional
    display 16
    effect 16
    vision 15
tif 108
tiff 108
time management 286
timetable 292
Titchener, Edward Bradford 24
title face 62
TM, trademark 272
tonal values 188
trackbacks 117
tracking 129
    codes 229
    system 229

trademark 272
    EU 274
    international 275
    national (German) 273
trademark law, US 273
trademark rights 272
traffic, information 229
transitionals 65, 74
Transmission Control Protocol/Internet
Protocol 114
trapping 176
tree structure 144, 300
triangle 27
trim 200
trimming 200
    errors 200
    marks 177
TrueType 90
Tschichold, Jan 42, 69, 81, 82, 84
ttf 108
TV formats 119
    HD 120, 121
    NTSC 119, 120, 123
    PAL 119, 120, 123
    SD 119
    SECAM 119
txt 108
type
    digital 66, 90
    family 70
    management 93
    movable 64
    origins of 63
    systems 72
type area 43, 81, 82
typeface 70
    characteristic style 62
    display faces 62
    headline typeface 62
    level of response 62
    reticent typeface 63

typeface and expression 61
  web design 134
typeface classification 72
  according to DIN 16 518 74
  according to Vox 73
typeface design 84
  basic rules of typeface design 85
Typeface Protection Act 271
typeline 68
typesetting 75
typesetting system, optomechanical 66
type size 57, 59
  reference size 59
  text size 59
types of plan 292
  flow plan 292
  Gantt method 293
  network plan 292
  PERT method 292
typographical point 57
typography 56
  macrotypography 56
  microtypography 56
  monitor typography 94

U
UCC, Universal Copyright Convention 255, 279
UCR, Undercolor Removal 178
UML 126
uncoated paper 161, 166, 170
unicode character coding 91
Uniform Resource Identifier 117
unique sales argument 222
unique selling proposition 223
United States Patent and Trademark Office 273, 279
Univers 66
Universal Copyright Convention 255, 279
uppercase 64, 67, 71
URI, Uniform Resource Identifier 117

usability 142
use cases prototypes 126
user
  behaviour 127
  scenarios 126
USP, unique selling proposition 223
  constructed 223
  natural 223
US PTO, United States Patent and Trademark Office 274, 279
utility contract 266
UV
  inks 187
  printing 190
  radiation 8
  varnish 197

V
value-added tax 247
vanishing point 18
varnish 197
  effect varnish 197
  emulsion varnish 198
  fragrant varnish 198
  UV varnish 197
  water-based varnish 198
varnishing 197
  shadow varnishing 197
  spot varnishing 197
Vasarely, Victor 33
VAT, value-added tax 247
VC-1 124
vector
  formats 110
  graphic 173
Venetian Old Style 74
viewpoint 20
Vinci, Leonardo da 15, 18, 19
Virtual Private Network 115
vision, three-dimensional 15
Visual Jockey 127

visual media manager 227
VJ, Visual Jockey 127
VJ-ing 127
Vox, Maximilien 73
VPN, Virtual Private Network 115

W
W3C, World Wide Web Consortium 139,
142, 146
Wahlbaum, Erich 65
WAN, Wide Area Network 114
warm light 10
waterless offset printing 190
watermark
    genuine 157
    non-genuine 158
    semi-genuine 158
wav 108
waves, electromagnetic 8
WCAG, Web Content Accessibility
Guidelines 142
Web 2.0 116
web browser 115, 124, 135
Web Content Accessibility Guidelines 142
web crawler 228
web design 128, 133, 146
    page size 133
    rules 133
    typefaces 134
web page 133
website 115, 133, 135, 141
    maintaining a 133
    setting up a 135
web standards 14
wedge
    colour wedge 188
    control wedge 182
    grey wedge 188
    screen wedge 189
wet beating 157
Wide Area Network 114

Wide Skyscraper 225
wide web 159
widow 83
Windows Bitmap 104
WIPO, World Intellectual Property
Organisation 255, 275, 279
wire-o binding 201
wire binding 201
WLAN, Wireless Local Area Network 114
wmf 108
WMV, Windows Media Video 124
WMV HD 124
woodcut 183
word space 75
works made for hire 259
World Intellectual Property Organisation
255, 275, 279
World Wide Web 115
World Wide Web Consortium 139, 146
worm's-eye view 19
writing conventions 79
WYSIWYG editors 135, 136

X
x-height 68
XHTML 136, 139
xls 108
xlsx 108
XML, Extensible Markup Language 139

Y
yellow 8

Z
zigzag folding 199
Zimbardo, Philip 31
zip 108
Zuse, Konrad 112

The Little Know-It-All

Common Sense for Designers

Edited by Robert Klanten, Mika Mischler and Silja Bilz
Text by Silja Bilz
Chapter introductions by Sonja Commentz, Simpeltext
"Color Management" text by Jan-Peter Homann, www.colormanagement.de
All text translated by Michael Robinson
Chapter Law translated by Kenneth Mills, Jean-Jacques Petrucci and Michael Robinson

Cover illustrations by Mika Mischler for dgv
Layout and illustration by Mika Mischler and Sabrina Grill for dgv
Production management by Janni Milstrey, Helga Beck and Vinzenz Geppert for dgv
Production assistance by Michael Chudoba for dgv

Proofreading by English Express GbR
Printed by Karl Grammlich GmbH, Pliezhausen, Germany

Set with the type system Compatil (Compatil Letter),
with the generous support of Linotype GmbH.

Die Gestalten Verlag, Berlin 2007

ISBN: 978-3-89955-167-9

Bibliographic information published by the Deutsche Nationalbibliothek
The Deutsche Nationalbibliothek lists this publication in the Deutsche Nationalbiblio-
grafie; detailed bibliographic data is available on the internet at http://dnb.ddb.de.

Dieses Buch ist als „Der kleine Besserwisser" in deutscher Sprache erschienen.
ISBN: 978-3-89955-166-2

Further information at: www.die-gestalten.de
Respect copyright, encourage creativity!

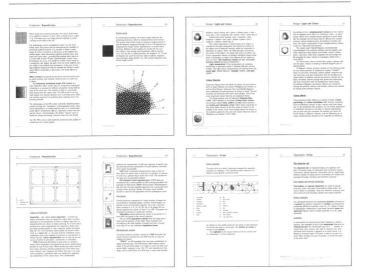

The Little Know-It-All provides the fundamental information designers need to know to thrive in their growing field of practice. It is an indispensable manual of the knowledge currently required of designers as the evolution of media redefines their role and expands the disciplines in which they must be competent.

Divided into the seven chapters "Design", "Typography", "Digital Media", "Production", "Marketing", "Law" and "Organisation", the book offers concise analysis as well as explanations of unique vocabulary. Written for a global audience, it expounds on various international formats and legalities.

With its thematic structure and resourceful index, The Little Know-It-All is a clever and comprehensive collection of essential practical information. Complete with graphics and illustrations supplementing the texts, it is both a stimulating reference book for students and newcomers and a trusty companion for design and media professionals to use in their everyday work.

Die Gestalten Verlag / ISBN: 978-3-89955-167-9

We would like to thank the following for their expert advice and support:

Chapter Design

Dietrich Bilz

Chapter Typography

Bruno Steinert, Geschäftsführer a.D. Linotype GmbH

Chapter Production

Daniel Grammlich, Axel Raidt, Ralf Fischer, Jan-Peter Homann, Mark Webster,
Karin Augustat, Martin Bretschneider, Vinzenz Geppert

Chapter Digital Media

Kora Kimpel, Franziska Schwarz, Tanja Diezmann, Joachim Sauter, Sven Haeusler,
Max Winde, Martin Steinröder, Lars Borgfeld, Sven Züge

Chapter Organisation

Stefan Trummer

Chapter Law

Evelyn Lüchter, Jens Fischer, Jean-Jacques Petrucci / Rechtsanwälte
Zimmermann und Decker, Hamburg;
Dominic Free / Forbes Anderson Free, London

Manuscript and Indexing

Bettina Fortak

Printed on MultiDesign by Classen-Papier

CLASSEN-PAPIER

Classen-Papier GmbH & Co.KG
Postfach 18 51 15
45201 Essen, Germany
Telephone +49.180.52 52 773 (12 cent/min within Germany)
Telefax +49.180.105 11 04
servicecenter@classen-papier.de